I0493815

Disclaimer

Book Title: Assessment of Factors Affecting Fire Performance of Mattresses: A Review

Book Author: Shonali Nazare; Rick D. Davis;

Book Abstract: An in-depth analysis of U.S. residential fire statistics shows that although the total number of fires and deaths due to mattress fires has dropped as a result of several regulatory approaches, the number of deaths per 1000 mattress/bedding fires has increased over past 3 decades. To address the increasing number of deaths due to mattress fires, the open flame mattress flammability regulation (16 CFR 1633) was recently introduced in the U.S. The 16 CFR 1633 prescribes performance standards rather than design standards; this allows manufacturers the flexibility to meet the needs of the consumer without sacrificing fire safety. This flammability regulation for residential mattress has generated much interest in understanding the burning behavior of mattresses as well as in developing new materials for mattress construction. To comply with this regulation, it is essential to understand mattress construction, fire performance testing, factors affecting mattress flammability, and compliance solutions. This report reviews the impact of current mattress flammability standards, examines factors affecting mattress flammability, and reviews full-scale and bench-scale test methods that are being developed for mattresses. The soft materials used in the mattress set, including cushioning materials, fire blocking materials, and tickings, act both individually and collectively to affect the fire performance. The performance of fire barrier materials designed to protect the inner cushioning material from heat and flame is largely dependent on the choice of cushioning material and ticking. When used with an incompatible combination of filling material and ticking, a fire barrier may fail to protect thermal degradation and subsequent burning of filling material. Some of the challenges in designing mattresses have been identified and reported here.

Citation: NIST TN - 1740

Keywords: Soft furnishings, mattresses; flammability standards; testing; heat release rate; ticking; barrier fabric; bedclothes.

NIST Technical Note 1740

Assessment of Factors Affecting Fire Performance of Mattresses: A Review

Shonali Nazaré
Rick. Davis

March 2012

NIST National Institute of Standards and Technology • U.S. Department of Commerce

NIST Technical Note 1740

Assessment of Factors Affecting Fire Performance of Mattress: A Review

Shonali Nazaré
Rick D. Davis

Fire Research Division
Engineering Laboratory
National Institute of Standards and Technology

March 2012

U.S. Department of Commerce
Gary Locke, Secretary

National Institute of Standards and Technology
Patrick D. Gallagher, Under Secretary of Commerce for Standards and Technology and Director

National Institute of Standards and Technology Technical Note
Natl. Inst. Stand. Techn. Techn. Report XXX, 53 pages (March 2012)
CODEN: NTNUE2

Abstract

An in-depth analysis of U.S. residential fire statistics shows that although the total number of fires and deaths due to mattress fires has dropped as a result of several regulatory approaches including reduced risk of ignition through reduced ignition propensity cigarettes, reducing fire spread through residential sprinklers, and reducing the inherent fire hazard of fuel sources through lower heat release (HR) mattresses, the number of deaths per 1000 mattress/bedding fires has increased over past 3 decades. To address the increasing number of deaths due to mattress fires, the open flame mattress flammability regulation (16 CFR 1633) was recently introduced in the U.S. The 16 CFR 1633 prescribes performance standards rather than design standards; this allows manufacturers the flexibility to meet the needs of the consumer without sacrificing fire safety. This flammability regulation for residential mattress has generated much interest in understanding the burning behavior of mattresses as well as in developing new materials for mattress construction. To comply with this regulation, it is essential to understand mattress construction, fire performance testing, factors affecting mattress flammability, and compliance solutions.

This report reviews the impact of current mattress flammability standards, examines factors affecting mattress flammability, and reviews full-scale and bench-scale test methods that are being developed for mattresses. The construction type, geometry, and size of a mattress are major factors in determining the fire threat of a mattress. The size effect is only significant for standard mattresses without any FR (flame retardant) modification. The soft materials used in the mattress set, including cushioning materials, fire blocking materials, and tickings, act both individually and collectively to affect the fire performance. The performance of fire barrier materials designed to protect the inner cushioning material from heat and flame is largely dependent on the choice of cushioning material and ticking. When used with an incompatible combination of filling material and ticking, a fire barrier may fail to protect thermal degradation and subsequent burning of filling material. Some of the challenges in designing mattresses have been identified and reported here.

Keywords

Soft furnishings, mattresses; flammability standards; testing; heat release rate; ticking; barrier fabric; bedclothes.

This page left intentionally blank.

Contents

List of Tables

List of Figures

List of Acronyms

ASTM	ASTM International
BHFTI	Bureau of Electronic and Appliance Repair, Home Furnishings and Thermal Insulation
Cal TB	California Technical Bulletin
CFR	Code of Federal Regulation
CPSC	Consumer Product Safety Commission
EPD	Environmental Product Development
EHS	Environment, Health, and Safety
FR	Flame Retardant
FIGRA	Fire growth rate index
HR	Heat Release
HRR	Heat Release Rate
IMO	International Marietime Organisation
NIST	National Institute of Standards and Technology
PHRR	Peak heat release rate
PUF	polyurethane foam
REACH	Registration, Evaluation, Authorization and Restriction of Chemicals
THR	Total heat released
TTP	Time to peak heat release

This page left intentionally blank.

1. Introduction

Over the past three decades, the landscape of mattress-related fires has significantly changed. In the U.S., numbers of mattress-related residential fires have consistently fallen with time. The credit for this improvement can be attributed in part to three factors: the introduction of smoldering ignition resistant mattresses regulated by 16 CFR 1632 [1], the introduction of reduced ignition propensity cigarettes, and the reduction of fire spread through residential sprinklers [2]. Despite these regulatory approaches, mattresses/beddings are still reported as the first items ignited in the majority of residential fires [3,4].

The fire threat of a mattress is determined by the propensity of component materials to ignite, the intensity with which they burn and the rate at which flames spread. The cover fabric, which is often termed as ticking in the mattress industry, is a mattress component that can char, melt or catch fire when in contact with an ignition source, for example, a smoldering cigarette or a small match flame. If the ticking forms a smoldering char, a considerable amount of heat may accumulate over a period of time and subsequently spread into the filling material. The ticking may also melt away from the ignition source, thereby exposing the underlying cushioning foam or filling. Smoldering may result in one of two possible outcomes as the heat penetrates into the filling: oxygen depletion may reduce the intensity of the ignition source and thereby extinguish the fire, or the filling material can ignite and fire begins to spread over the mattress assembly. Once the fire becomes an open flame, the bedclothes may catch fire and act as a high intensity secondary ignition source, leading to ignition of the underlying materials, possible increase the threat of fire spread to other items in the room [5]. In addition to the direct threat of ignition, the hot gases and smoke from the burning bed assembly will accumulate under the ceiling. The temperature of this hot layer is very high, and its radiant heat can eventually ignite all flammable items in the room, leading to room flashover. Flashover typically occurs when the heat release from a burning bedding assembly exceeds 1000 kW [6]. A fire of this size in a confined space results in rapid generation of carbon monoxide, which poses another serious threat to occupants elsewhere in the building. Thus, a bedding fire that begins with a smoldering cigarette or the open flame of a match has the potential to translate into a large fire with serious consequences.

New materials, constructions, and designs have been developed to meet the consumer's changing comfort and aesthetic needs while also addressing more rigorous flammability requirements from federal regulatory agencies. Mattresses are complex products that are used by human beings for a long period of time during restful sleep. Generally the function of a residential mattress is to provide comfortable surface to rest or sleep. Mattresses used in public occupancies often have specific purposes and their functional requirements can be very different from the residential mattresses. Specific functions can be fulfilled with significant variations in construction geometries, support materials, cushioning materials, and textiles. These variations have significant impacts on the flammability properties of a mattress. Another complicating factor is the variability in the composition of these components and their assembly that may be introduced during manufacturing. In addition, the development of a fire is sensitive to the composition and geometric arrangement of the fuel. Manufacturing variability can therefore not only significantly impact the fire behaviour of each component alone, but can also change the synergistic or antagonistic interactions of the components with each other. The interaction of all components in the final mattress set (mattress with foundation) is what ultimately determines the fire threat.

The presence of bedclothes including sheets, blankets, bedspreads, pillows, and bed valances (also known as protective skirts) could also dictate or overwhelm mattress component interactions and hence the fire threat from mattresses.

This report aims to the review the impact of current mattress flammability standards, examine factors affecting mattress flammability and review mattress flammability test methods and compliance solutions. This review is timely in light of the newly introduced flammability standard for mattresses, and new materials and construction styles used to comply with this flammability standard.

2. Mattress Flammability Regulations and their Impact on Fire Statistics

The majority of national and international mattress flammability standards and test methods listed in Table 1 are applicable to mattresses used in high occupancy public buildings. It is only in developed countries (the U.S., Canada, UK, France and Norway) that residential mattresses have to comply with flammability regulations. In the United Kingdom the BS 6807 [7] standard, now replaced by BS EN 597 [8], is used to assess the basic ignitability properties of mattresses and foundations. Most European countries use EN 597 standard to evaluate ignitability of a mattresses. For mattresses used in high occupancy public buildings, for example in hotels, hospitals, and other public places, the BS 7177 [9] standard specifies various combinations of ignition sources to represent four different hazard classifications as low, medium, high, or very high. The Canadian mattress flammability test (CAN2-4.2-M77 [10]) is unique in that this is only the small-scale mock-up test to determine smoldering ignition resistance of a mattress and uses single lighted cigarette as smoldering ignition source. Generally, mattress flammability standards prescribe full-scale testing of a prototype mattress (mattress without bedclothes) when exposed to different ignition sources. The ignition sources defined for the mattress flammability test in Table 1 represent the fire hazard of a mattress. For example, the Michigan Roll-up test defined by the Boston Fire Department, U.S., requires testing of mattresses used in jails to be tested as rolled up mattresses stuffed with newspaper. This configuration of mattress and ignition source (burning newspaper) represents a fire hazard stemming from a representative, deliberate act.

Besides, there are only few test methods developed to evaluate burning behavior of mattress components such as textile thermal barrier materials (ASTM D 7140), sewing thread, tape and edge components (ASTM D 7016). Germany has classification scheme for bedding components based on DIN EN 14533 [11] whereas NT FIRE 037 (Table 1) determines ignitability of bedclothes including mattress pads. Standards and test methods listed in Table 1 may be mandatory or voluntary. Mandatory standards, also known as regulatory standards, are incorporated into government regulations with which products must comply. Voluntary standards are often used for quality control in industry or for development of new products.

The fire performance of mattresses is regulated in the United States according to the type of ignition source, either smoldering or opens flame. The smoldering ignition test measures the char length over the mattress surface and the extent of damage to the mattress after a specified time period (Table 1). Open flame ignition tests measure the heat release rate, total heat released, and/or mass loss for the burning mattress set during a specified time period (Table 1).

The current U.S. mattress flammability standards that have the most significant impact on industry and the consumer are 16 CFR 1632 [1] and 16 CFR 1633[12]. These flammability standards for residential mattresses are derived from test methods developed for mattresses used in public occupancies (Cal TB 129). The 16 CFR 1632 regulation, introduced by the Consumer Product Safety Commission (CPSC) in 1973, defines the fire resistance of mattresses to cigarette ignition, a smoldering source. The 16 CFR 1633 regulation, introduced in 2007, defines the resistance of mattresses to open flame ignition sources. These two federal flammability regulatory standards are mandatory, and all manufacturers must comply with them in order to sell residential mattresses in the U.S. (Table 1).

Across the world, very few comprehensive statistics exist, especially those which attempt to relate deaths and injuries to first items of ignition in residential buildings. International reporting of fire statistics is not standardized and no common international basis exists for the gathering and interpreting of such information. Systematic fire incidence reporting system is now well established in the US, UK and Canada and by far shows residential/furnishing fires to be significant cause of fire deaths. This section however, focuses on U.S. mattress flammability regulations and their impact on fire statistics.

A significant change in the U.S. residential fire losses related to mattresses/beddings as the first item of ignition was observed beginning in the early 1980s, almost ten years after the introduction of the smoldering ignition performance regulation (16 CFR 1632) (Figure 1a to Figure 1c)[13]. Since this is consistent with the time scale over which residential mattresses are replaced, many experts agree that the 30 % decrease in the number of residential mattress/bedding-related fires due to smoking materials during the early 1980's (Figure 1a) was primarily driven by 16 CFR 1632 [2]. The continuing decrease in the number of such fires over the next two decades can be attributed at least in part to this regulation. However, other changes over this period, including the commercialization of low smoldering cigarettes and the reporting methodology for generating fire statistics, make it difficult to separate out its specific contribution. From Figure 1b and Figure 1c, it appears that the smoldering ignition flammability regulation may have had a lower immediate impact on civilian fire fatalities and injuries [2]. By 2006, however, the combination of 16 CFR 1632, low smoldering cigarettes, and other factors had significantly reduced all such losses – by 93 % in the number of fires and 73 % and 68 % in the number of civilian injuries and deaths respectively. For open flame ignition, the number of residential mattress/bedding-related fires decreased significantly between 1980 and 2006, while the average number of civilian deaths and injuries showed a downward trend with large year-to-year fluctuations.

The U.S. fire statistics [14] for all residential fires include a breakdown of ignition sources, permitting fires from smoldering sources (cigarettes) to be distinguished from those from open flame sources (matches, lighters, and candles). The incidence reporting system [14] however does not differentiate between bedding and mattress fires. Figure 2a shows that the number of residential fires caused by smoldering cigarettes has decreased much more than the number of fires caused by flaming ignition sources. The number of personal injuries caused by open flame ignition sources has decreased by less than those caused due to smoldering ignition sources (Figure 2b). However, deaths from cigarettes continue to be a factor of 2 higher (Figure 2c).

In 1994, for the first time since fire losses have been tracked at this level of detail, there were more mattress-bedding fires caused by open flame ignition sources than by smoking materials (Figures 1a and 2a). Other studies have also shown an increasing trend towards open flame bedding fires [15,16]. Open flame ignition represents a more immediate hazard than smoldering ignition. Smoldering fires on mattress may take 25 min or longer to transition into flames, the point at which mattress fires ignited by open flames begin [6]. Open flame fires provide a short time window for detection, escape and fire response. Moreover, despite the drop in the total number of fires and deaths due to mattress fires [13], deaths per 1000 mattress/bedding fires have increased with time (Figure 3). This was one of the motivations for CPSC to introduce a sister regulation to 16 CFR 1632, which defined the open flame performance of mattresses (16 CFR 1633). All mattresses brought into the U.S. market since 2007 have been required to comply with 16 CFR 1633. While it is expected that 16 CFR 1633 will significantly reduce fire losses, the true impact of this standard is not expected to be realized for at least another 5 to 10 years, after a majority of old non-compliant mattresses are replaced with new compliant ones [17]. Increase in number of deaths per 1000 mattress/bedding fires could also be due to increased fire hazard of bedclothes. Bedclothes flammability studies [18,19,20] have shown that bedclothes have the potential to result in flashover in a few minutes after ignition. Formal regulation of flammability of bedclothes (Cal TB 604, see Table 1) was proposed by the Bureau of Electronic and Appliance Repair, Home Furnishings and Thermal Insulation (BHFTI) in the state of California, but this is now suspended. The Canadian Underwriters' Laboratories (UL) developed test methods for mattresses with bedclothes (UL 2060) which is also withdrawn.

Both 16 CFR 1632 and 16 CFR 1633 are performance standards rather than design standards. They do not address specific approaches for achieving compliance (such as through barrier materials or fire retardants); instead, they describe how to conduct the tests and provide pass/fail criteria. This allows manufacturers the flexibility to meet the needs of the consumer without sacrificing fire safety. To comply with these standards, it is essential to understand mattress construction, fire performance testing, factors affecting mattress flammability, and compliance solutions. These topics are discussed in the following sections.

3. Mattress Construction and Component Materials

The sizes, geometry, and construction of mattress sets are reasonably well standardized (Table 2) and hence can be assigned a specific number. The soft components of a mattress are manufacturer-dependent, with many highly engineered combinations of fibers, fabrics and foam available. Excluding ticking variations, mattresses are available in several thousand variations of design and construction (Table 2). The primary components of mattresses are described below.

3.1. Frame and foundation

A typical mattress set consists of three components: the frame, foundation and mattress (Figure 4). The frame is the support for the mattress set and is usually constructed from wood or metal. The presence or absence of the frame may affect the fire threat of a burning mattress. For example, a pool fire generated from flaming molten drips of burning bedclothes can result in rapid flame spread under the mattress, but this flame spread from under a mattress may not occur if the mattress set rests directly on the floor [18].

The purpose of a foundation is to provide support to the mattress, enhance mattress performance, and extend the service life of a mattress. The foundation and mattress are generally paired as a matched set. Using a foundation that is not well-matched with the mattress can decrease mattress performance and service life, and may also affect the fire performance of the mattress set. The most common type of foundation is a box spring (Figure 4), which is constructed of several springs or shock absorbing torsion modules mounted on a rigid metal support. The boxspring is covered by a ticking, which usually matches the ticking fabric of a mattress. Foundations with foam filling and cotton battings have also been reported but are becoming obsolete in modern mattress sets.

3.2. Mattress support system

Mattresses are classified by the type of support system: innerspring, solid foam, cotton batting, air, or water. Air and water mattresses account for less than 20 % of the U.S. market, solid foam (viscoelastic) mattresses and innerspring mattresses account for nearly all of the remaining 80 % of the U.S. market. The focus of this report is mainly on innerspring mattresses as there is very little information on the flammability of solid foam (viscoelastic) mattresses. An innerspring mattress is constructed from metal or plastic springs that may be separately housed in individual fabric sheaths or attached to a metal frame (Figure 4a). For single-sided mattresses (also known as 'no flip' mattresses), the innerspring is covered by a comfort layer on one side and a low-cost non-woven backing material on the other. A comfort layer covers both sides for double-sided mattresses.

3.3. Cushioning and comfort layer

The comfort layer is divided into three subcategories: the quilt, the insulator layer, and the cushioning layer. The quilt is the top layer of the mattress. It provides a soft surface texture and a level of firmness that can be varied by changing the material and the details of construction. The quilt consists of the ticking plus a low-density foam or fiber batting that is stitched to its underside. These two layers are sewn to a tape edge that attaches to the border quilting around the perimeter of the mattress. The insulating layer conserves the warmth of a sleeping person, and can be a fiber batting or layers of non-woven fabrics. The cushioning layer provides an extra layer of comfort, and may include flat or convoluted PUF, shredded pads of compressed polyester, or fiber battings. The insulator and cushioning layers can be stacked in varying sequences between the quilt and the innerspring support. With the introduction of 16 CFR 1633 in 2007, most mattress manufacturers changed to single-sided mattresses because of the expense of complying with flammability regulations for both sides of the bedding [21].

3.4. Fire blocking materials

The purpose of fire blocking materials is to reduce the flammability of soft furnishings by preventing or delaying direct flame impingement and heat transfer from the flames to the core cushioning components. A variety of fire blocking technologies using various types of fabrics and fibers has been developed. A detailed review is available that covers fire blocking mechanisms and technologies used in soft furnishings in general [22]. In this section, fire blocking materials that are specifically used for mattress applications are discussed.

Fire blocking materials were commonly used in institutional mattresses that are required to pass the open flame ignition resistance test. However with the introduction of the 16 CFR 1633 regulation, the immediate response to comply with the regulation was to introduce fire blocking materials in residential mattress construction.

Institutional mattresses use both active and passive fire blocking technologies. A passive fire barrier is made from inherently fire resistant fibers. It serves as a physical and/or thermal barrier between some or all of the fuel and the potential ignition source. Glass fiber battings or woven glass fiber fabrics are commonly used in institutional mattresses, although fiberglass flame barriers have the disadvantages of poor durability (due to glass-to-glass abrasion) and lack of resiliency. Active fire blocking can be achieved through a fire retardant (FR) coating on a glass fiber substrate. These barrier materials have a chemical effect on the fire. They can alter the pyrolysis process to reduce the amount of flammable volatiles and suppress the flames from the ignition source, prevent the ticking from burning, and prevent the ignition of interior cushioning material.

Fire barriers are not recommended in public occupancies that have a relatively high risk of vandalism, such as prisons and mental hospitals. In public institutions with high risk occupancies, densified polyester batting is often used as a filling material instead of highly flammable foam. Densified polyester batting is difficult to ignite as the thermoplastic polyester melts and shrinks away from the ignition source thereby making it difficult to ignite a mattress.

Unlike institutional mattresses, comfort and aesthetics are of primary importance in the case of residential mattresses. In residential settings, therefore, fire performance must be achieved while still maintaining both comfort and aesthetics. For this reason, nonwoven, highloft battings are more commonly used as fire barriers in residential mattresses. Nonwoven cotton battings treated with boric acid have been used for many years as fire barriers in mattresses [23]. However, boric acid treatment may have problems associated with chalking, color change and undesirable texture [24]. Highloft battings of FR rayon blended with polyester fibers have gained popularity especially after the introduction of 16 CFR 1633. These barrier materials are viewed as an environmentally friendly and economically practical approach to comply with 16 CFR 1633.

Another fire blocking technology uses core spun yarn to produce knitted barrier materials [25]. In these designs, inherently fire resistant glass fiber forms the core, which is coated with a blend of char forming FR fiber and polyester fiber. Polyester fiber is primarily responsible for its aesthetic and comfort properties. The thermally stable core maintains the structural integrity during a fire by providing a woven framework (grid) for the char layer (lattice) formed by the thermal decomposition of the sheath fiber while burning. The composition of the core and sheath can be tailored to satisfy barrier performance and comfort requirements.

3.5. Ticking

Current residential mattresses use a wide range of tickings, including pile fabrics, knits, woven fabrics and jacquard designs. To address issues of physiological comfort, fire safety, and the growing incidence of allergies within the U.S. population, a variety of functional coatings, including water-proof, anti-bacterial, anti-fungal, and/or flame retardant finishes, are applied to the ticking of the mattress. The majority of modern ticking materials have a high polypropylene and/or polyester fiber count, with the fiber content varying significantly with the fabric structure and design pattern. While cotton, polyester and polypropylene fibers dominate the ticking industry, blends of luxury fibers, such as wool and silk, are becoming more prevalent. Renewable resources like corn, soybean and bamboo fibers are also gaining popularity as more environmentally friendly alternatives. Viscose rayon derived from bamboo is of particularly high interest because of its inherent anti-bacterial and anti-fungal properties and its good breathability and moisture absorption. However, very little is known about the flammability of these green alternatives.

For institutional mattresses, fire performance takes precedence over comfort and aesthetics. Polyvinyl tickings and fiberglass substrates with FR coatings are the preferred choices for institutional mattresses. A typical FR coating formulation consists of FRs (typically gas phase FRs), fillers, synergists and application ancillaries (e.g., polymeric resin binder, fabric softeners, and cross linking agents). A halogen-containing polymer, combined with vinyl fluoride and finely dispersed antimony oxide is commonly used for coating ticking employed in heavy use applications such as healthcare mattresses.

4. Mattress Flammability Testing

4.1. Full-scale testing

Both mattress flammability standards, 16 CFR 1632 and 16 CFR 1633, require full scale testing of all prototype mattresses or mattress sets introduced for sale in the United States. In order to minimize the testing burden on the manufacturer, a representative mattress or mattress set to be placed on the market can be tested. If this sample meets both cigarette ignition resistance and open flame ignition resistance test criteria, it then becomes a 'qualified prototype' that can be used as a model for the production of these mattresses, as long as the materials, components, design, and method of assembly remain unchanged. Furthermore, manufacturers have been granted the flexibility to produce similar mattresses of differing sizes and to use materials, components, and methods of assembly whose FR performance is similar to or better than the qualified prototype. Such 'subordinate prototypes' do not require additional testing as long as a record of the manufacturing specification and a description of the variation from the 'qualified prototype' are available. The manufacturer is also required to show sufficient documented evidence that the changes in the subordinate prototype will not cause the prototype to exceed the specified test criteria. There also exists a possibility of 'pooling' the qualified prototypes, whereby two or more manufacturers can use qualified prototypes to produce mattress sets. When using pooled prototypes, manufacturers are required to conduct one successful confirmation test.

The fire behavior aspects that are generally examined for an open flame ignition test include the heat release rate (HRR) as a function of time, the time and level of the peak heat release rate (PHRR), the total heat released (THR), the rate of flame spread, and the mass loss. Mattresses or bedding assemblies are placed on top of a large load cell during the flammability test to measure sample weight as a function of time. The test method described in 16 CFR 1633 uses dual T-shaped propane burners with a heat flux of 65 kW/m^2 and 45 kW/m^2 for top and side burners respectively. The top surface of the mattress is exposed to the burner for 70 s, and the side is exposed for 50 s. The test criteria for the first 10 min of test duration are that the THR shall not exceed 15 MJ and that the PHRR throughout the test (30 min) shall not exceed 200 kW. Heat release rate is measured by oxygen consumption calorimetry, either in an open hood environment or inside a room.

The two types of environment used in fire testing, open hood and room, may result in very different fire behaviors. Ohlemiller [26] studied fire tests in both environments and identified two mechanisms by which a room environment could affect the burning behavior of a mattress: through thermal feedback from the smoke layer to the burning surfaces and through oxygen limitation, which depends on the openings in the room and their effect on ventilation. This susceptibility to room effects makes it difficult to achieve inter-laboratory agreement on data and on the evaluation of the fire hazard for mattresses of similar construction. It is therefore important to develop bench-scale tests that have good reproducibility and good correlation with full-scale test data.

4.2. Bench-scale testing

In an early (1981) attempt toward bench scale fire tests for mattresses, Babrauskas [27] developed bench scale test procedures for classification of mattress burning behavior when exposed to a flaming ignition source, and compared the results to full-scale testing. Heat release rate and smoke production were identified as parameters that enable performance classification of mattresses by both full-scale and bench-scale test procedures. However, flame spread and ignition properties measured using bench-scale test protocols failed to characterize mattress behavior consonant with full scale tests.

In a later study, Ohlemiller [26] examined the feasibility of developing a bench-scale protocol for possible use in CFR 1633 compliance testing of commercial mattress designs. The goal was to design a bench-scale method to predict the immediate response of a mattress or mattress set to a gas burner ignition source. The test specimen for the proposed bench scale test procedure was an abbreviated composite of the mattress or mattress/foundation assembly, and the measured parameter was the time to burn through the specimen. The bench scale test procedure had its own limitations with respect to sample preparation, reproducibility of test data, and did not fully reproduce burning behavior in a full scale 16 CFR 1633 type test. The test was therefore considered ineffective and insufficient. Moreover, since the test required specially constructed samples and every failure mode, for example, determination of whether the burn-through occurs over the mattress surface or at seams required a separate test, the whole approach was regarded to be impractical and uneconomical.

The future bench scale testing methods should be based on simple but scientifically sound principles that may be employed for screening of materials. Bench scale flammability tests are useful in that several material fire properties can be derived and the data can be used for relative ranking of materials. Another potential approach is to use data from bench scale tests in mathematical models to predict large scale fire behavior. However, at the present time predictive testing has too many unquantifiable variables, and so it will likely remain a research tool in the near-term future.

5. Factors Influencing Fire Performance of Mattresses

The fire performance of a mattress depends on each of the components described in Section 3, along with the possible synergisms or antagonisms that may exist among them [16].

5.1. Mattress dimensions

The construction and geometry of a mattress and foundation can be major factors affecting the fire performance of a mattress set. The fit of the foundation to the frame, the presence of the foundation and bedclothes all contribute to the fire hazard of a mattress.

Mismatch between foundation and frame

The geometry of the foundation is especially important when the foundation is placed on a metal or a wooden frame. If the foundation does not fit precisely within the supporting frame, the small gap between the frame and the foundation offers a potential path for small flames on the foundation ticking to reach the underside of the foundation [26]. If the underside of the foundation is not protected by fire barrier materials, the flames could then easily reach the more flammable materials used in mattress construction and the fire can result in flashover in a matter of a few minutes after ignition. To overcome this problem, the recent 16 CFR 1633 regulation specifies that the bed frame must match the dimensions of the mattress set.

Impact of mattress size

Although the 16 CFR 1633 regulation does not specify mattress size, a twin mattress is typically used in testing, since the fuel load and manufacturing cost are significantly less for a twin mattress than for a queen or king size mattress. Ohlemiller [6,18] studied the dependence of the PHRR of a bedding assembly (mattress set with bedclothes) on the size of the mattress and on whether or not a fire blocking barrier or a FR ticking is used. The mattress sets used in this study were standard innerspring mattresses with box spring foundations. The bedclothes consisted of a mattress pad (polyester/cotton batting), fitted and flat sheets (50:50 polyester: cotton), a blanket (100 % acrylic), a comforter and a pillow (100 % polyester fiberfill encased within a polyester/cotton shell) and a pillowcase (50:50 polyester:cotton). At 2293 kW ± 25 kW, the PHRR of a twin bed that contained a PUF was ≈ 36% less than the PHRR of a king bed of the same construction (3610 kW ± 339 kW) (set M-I in Figure 5). This increase in PHRR for the king size bed is less than a factor of two even though the surface area is twice as large for the king size bed [18]. Similar sized mattress sets constructed with a fire blocking barrier fabric (set M-II in Figure 5) reduced the PHRR by an order of magnitude. The PHRR of the king and twin mattress sets (set M-II) were different within the standard uncertainty of the measurements. However, compared to set M-I where the PHRR for king size bed was significantly higher than the twin size bed, the values of PHRR for the king and twin mattress sets of M-II design were quite comparable. An FR modified mattress using a FR ticking (set M-III in Figure 5) was not as effective as one using a barrier fabric (set M-II) in reducing PHRR. This can be attributed to the failure of the FR ticking alone to protect underlying cushioning layers from burning bedclothes. Neither FR modified mattress set (M-II and M-III) showed a noticeable difference in PHRR between twin and king size beds. The study suggests that the size effect is only significant for standard mattresses without any FR modification.

Since the fire losses from a burning mattress depend not only on the size of the fire but also on how quickly it grows, the FIre Growth RAte (FIGRA) index [28] could be a more appropriate indicator of fire performance (inset of Figure 5). The fire growth rate index is calculated by dividing the peak heat release by time to peak heat release (FIGRA = PHRR/TPP), and can be used to estimate both the predicted fire spread rate and the fire hazard. The higher the FIGRA index value, the higher the fire hazard. Therefore, the FIGRA in reality becomes a heat acceleration parameter. However, care should be taken while predicting the fire threat of complete bedding assembly using FIGRA, since it has been shown that under certain conditions the HRR curves for these bedding assemblies show two distinct peaks [19]. In these cases, the first peak is dominated by bedclothes with little contribution from the mattress and the second is dominated by the mattress and foundation.

The inset in Figure 5 compares FIGRA index values (calculated using PHRR values from the first peak) for mattresses M-I, M-II and M-III. The FIGRA value of 10.87 for the king size bed reflects a greater fire hazard relative to the twin bed (FIGRA index of 5.50) with non-FR mattresses. For bedding assemblies with mattresses M-II and M-III, FIGRA values are very similar for twin and king sizes. Thus the comparison of bedding assemblies using PHRR provides similar results to that using FIGRA values.

5.2. Mattress Construction

Interaction of the mattress and foundation

It is possible for a foundation to improve the fire performance of a mattress set by reducing the air flow to the bottom of the mattress, thereby creating an oxygen-deprived environment that can slow down fire growth or result in self-extinguishment. If the foundation is constructed with flammable materials, however, the additional fuel can contribute towards the heat release of the entire mattress set. Peak heat release rates for open flame testing of various mattresses with and without foundations are provided in Figure 6. All mattresses used to compare the fire performance of different foundations in this study [29] used a similar innerspring construction with different PUF fillings. Based on their construction details and component attributes, mattresses A and D are classified by the authors as low hazard mattresses, whereas mattresses B and C are classified as high fire threat mattresses. The presence of a standard metal and wood foundation with ticking had little impact on the PHRR of innerspring mattresses with conventional PUF filling (set A in Figure 6) or with Cal TB 117 grade FR-PUF filling [30] (set D in Figure 6). Adding a foundation containing a cotton batting resulted in a nearly 50 % decrease in the PHRR for mattress set B compared to mattress B alone. With a PUF filler instead of a cotton batting in the foundation tested with mattress C (similar component materials and construction to mattress B) the PHRR nearly doubled (800 kW for set B and 1580 kW for set C). Unlike set B, the fire performance of mattress set C is slightly worse than for mattress C without the foundation.

Fires that begin in the foundation (foundation-forced fires) usually originate from the foundation side walls and eventually spread laterally onto the underside of the foundation top pad, with subsequent ignition of the wooden base [18]. The fire spread can ignite the mattress and can also aid in flame spread across the mattress or to other objects in the room. Many mattress fires resulting in flashovers have been attributed to foundation-forced fires. King size beds constructed by placing a king size mattress on top of two adjacent long twin-sized foundations generate an additional flammability concern stemming from the crevice between the two foundations under the longitudinal centerline of the mattress [18]. When a fire in the horizontal space between the mattress and the foundation reaches the vertical crevice between the two parts of the foundation, the flames spread inward into the vertical crevice over the full foundation height and move onto the bottom of the foundation. It has been speculated [18] that using a double-sided mattress may mitigate this fire hazard.

Innerspring vs. solid core constructions

Damant and Nurbakhsh [29] reported on comprehensive full-scale fire tests conducted on both residential and institutional mattress constructions. Mattresses were tested according to the Cal TB 121 [31] standard (Table 1), with the galvanized metal container with 10 double sheets of loosely wadded newspaper replaced by a T-shaped gas burner positioned parallel to the bottom horizontal surface of the mattress. Selected results and descriptions of mattress components and construction are given in Table 3. Comparative data for innerspring and solid foam core constructions with various filling components are graphically presented in Figure 7. The PHR of a solid foam core mattress with non-FR PUF is significantly greater than that of an innerspring mattress with similar filling material. Unless the PUF is FR modified, 100 % mass of a solid core mattress is combustible. The recent Canadian study on residential twin sized mattresses concluded that solid core mattresses with non-FR PUF have potential of causing flashover [32]. Mattresses with greater amounts of combustible materials have higher PHR and a higher FIGRA value (Table 4). However, the burning behavior of innerspring mattress filled with melamine type foam showed a significantly higher heat release (453 kW) than the solid core cellular foam mattress (39 kW) filled with similar melamine type foam. The higher PHR of the innerspring mattress filled with melamine type foam can be attributed to greater burning of the FR foam in a well-ventilated innerspring type of mattress construction. Data in Table 3 indicates a higher mass loss of 9.842 kg for an innerspring mattress filled with the melamine type foam compared to a minimal mass loss of 0.816 kg for the solid core analog. Thus, the fuel load of a mattress may not by itself be used to predict its fire performance. For mattresses incorporating Cal TB 117 grade foam [30] or boric acid treated cotton fillings in Figure 7, the type of construction (innerspring or solid core) had a minimal impact on peak heat release values.

5.3. Ticking

The direct contribution of the ticking to the fire threat is considered to be low because the heat release potential of the ticking is small compared to the mattress. In residential mattresses, ticking is considered as a sacrificial layer, which means it is designed to burn quickly and release very little heat. This is because the rapid burning of the ticking prevents the flames from staying in any one area sufficiently long to ignite the underlying components of the mattress. However, full scale open flame testing of residential mattresses over the last couple of years have shown that a change in the ticking alone has significantly increased the heat release rate of mattresses by more than 200 % in some cases (unpublished observations).

According to 16 CFR 1632 all ticking are classified according to their smoldering performance and defines the criteria for retesting the fire performance compliance of the mattress set when the ticking is changed (16 CFR 1632.6). As long as the ticking is replaced with a ticking of the same classification and nothing else is changed, then retesting the compliance of the mattress set is not required. A more detailed description of ticking classifications is provided in Table 5. Unlike 16 CFR 1632, 16 CFR 1633 does not define a ticking classification and does not require retesting to determine the open flame performance of a mattress set if only the ticking has been changed. This is primarily because the ticking was not found to significantly affect the open flame performance of mattresses tested at NIST [6, 19]. However, since the adoption of the regulation, field data has shown a significant increase in PHRR and THR of mattress sets (even to the extent of being non-compliant) in cases where only the ticking has been changed. This suggests that the ticking may not be sacrificial in these cases. The original experiments conducted at NIST [6, 19], which were used by CPSC to develop 16 CFR 1633, involved only a few ticking types and constructions that represented the majority of the market at that time. However, the number of ticking materials and construction types has significantly increased since 2007, and these changes may have been responsible for altering the fire performance of mattresses.

The impact of tickings with different fiber content and fabric finishes on the fire performance of mattresses has been studied [33] using the test procedure described in 16 CFR 1633. The results from this study are summarized in Table 6. All tested residential mattresses had a similar innerspring construction except for the tickings. The fire performance of these mattresses with varying tickings in terms of FIGRA is shown graphically in Figure 8. T-1 and T-2 samples have a THR of 12.7 MJ and 13.3 MJ, respectively, which are within 20 % of the 15 MJ THR failure criteria in 16 CFR 1633. However, these specimens have very low FIGRA values (0.08 and 0.06 respectively), which suggests that the estimated fire spread and size of the resulting fires may be significantly lower than for the specimens with lower THR values (T-3, T-4 and T-6 and T-7). Data in Table 6 suggest that changing the ticking component significantly alters the fire performance of the mattress. Previous studies [34] have also shown that PHRR values might be dependent upon the fabric surface and fabric construction of the tickings in addition to their fiber content and fabric finish.

Study [35] on the flammability testing of mattress composites have shown that the quilting also affects burning behavior, such that quilted specimens exhibit slightly higher THR values as compared to non-quilted specimens when tested under the cone calorimeter. One of the probable reasons for this kind of fire performance may be attributed to the fact that flame spread in quilted specimens is much slower. The quilting acts as flame arrestors and hence the material burns slowly but completely to give higher THR values.

Interaction with fire barrier materials
Tickings perform differently in the presence or absence of fire barrier materials. A large majority (about 80 %) of mattresses with a polyvinyl chloride (PVC) ticking pass the high occupancy dwelling open flame test (TB 129) without using a fire barrier material, because PVC tickings are active fire barriers that self extinguish. Approximately 20 % of the mattresses with PVC coated tickings fail due to antagonistic reactions of the highly plasticized PVC coated fabrics with other components of upholstery. Mattresses with cotton/fiberglass tickings also do not require an additional barrier fabric layer to protect the underlying cushioning layer. In this case, the cotton/fiberglass ticking acts as a passive fire barrier that physically prevents flame and heat transfer to the underlying cushioning layer. On the other hand, mattresses with cotton ticking do require a fire barrier in order to pass the open flame test. This is partly because cotton

is extremely flammable and cotton tickings burn with a high rate of flame spread, thereby exposing underlying cushioning layers to the open flames. A study [16] that investigated the effects of cover fabrics and filling materials with and without fire barriers showed that polyester and polyester blend ticking perform poorly in the presence of a barrier material. The role of barrier materials is discussed in much greater detail in the following section.

5.4. Fire barrier materials

Innerspring mattresses

The impact of fire barriers on the fire performance of innerspring mattresses with the same construction but different filling materials is shown in Figure 9. With a fire barrier (fibreglass fabric), these innerspring mattresses were able to pass with 100 % success the high occupancy dwelling, open flame ignition test for mattresses (Cal TB 129), regardless of the filling type (e.g., standard PUF, polyester fiber batting/PUF, or a cotton batting/ felt) [36]. Without the fire barrier fabric, the TB 129 performance of the mattresses was inconsistent, with the degree of failure depending on the type of filling material. For example, PUF innerspring mattresses had a success rate of 44 %, signifying four passes out of 10 tests. The cotton batting/ PUF and polyester fiber batting/cotton felt/PUF innerspring mattresses yielded a success rate two times greater at approximately 88 %. Innerspring mattresses with a polyester fiber batting combined with an insulator pad and PUF or cotton batting had a 100 % TB 129 success rate without the need for a BF.

Solid core mattresses

In this same study, the researchers determined that solid core mattresses passed TB 129 without using a BF. This is presumably a result of the restricted airflow in a solid core mattress, which restricts the entrainment of oxygen needed to sustain pyrolysis. This suggests that under the right construction and with the right combination of materials it may be possible to pass TB 129 without using a BF. However, this would not necessarily provide a product that is desirable by the manufacturer or consumer (e.g., it may not be comfortable, attractive, or cost-effective).

Institutional vs. residential mattresses

The materials and constructions discussed above for passing TB 129 are generally used for institutional mattresses. Ticking with polyester or polyester blends that are used in many residential mattresses behave very differently in the presence of fire barriers. The effects of melting and dripping can have a varied impact on the flammability of a mattress. Fire barriers often fail to protect the underlying material when melting and shrinkage occur. This can cause tensile stresses to develop within the barrier material, resulting in breakage that allows flame and heat to penetrate. This particular failure mechanism is a major concern for barriers based on charring organic fibers.

Compliance data for residential mattresses with highloft or other newly engineered fire barriers are currently not available. Several polyester blend tickings are being currently investigated and their fire performance with and without fire barrier materials is being studied in our Fire Research Division (FRD).

5.5. Bedclothes

Over the last 20 years there have been two sets of detailed studies by Damant and Nurbakhsh [29] and Ohlemiller [19] that demonstrate that burning bedclothes on a mattress have the potential to bring a room to flashover. Since both studies were performed before 2007, their conclusions were based on experiments conducted on mattresses that were not compliant with 16 CFR 1633. In three different scenarios using mattresses/foundation sets with and without bedclothes, Damant [29] reported that the presence of bedclothes (including a mattress pad, two bed sheets, a bed pillow with pillowcase, and two blankets) caused a 10 % to 30 % increase in the PHRR (Figure 10). In order to determine the specific contribution of bedclothes towards the fire performance of the bedding assembly with different mattress construction and uniform set of bedclothes, the heat release test data for mattresses with complete bedding assembly was normalized by subtracting the heat release data of bedclothes alone. The heat release data for the bedclothes shown in Figure 11 was determined by burning the bedclothes assembly on an "inert" mattress made of fiberglass insulation wrapped in a fiberglass barrier fabric. The bedclothes on the inert surface (black bar) were reported to generate PHRR of 146 kW, mass loss at 10 min of 2.4 kg, and ceiling temperature of 200 °C. The mattress constructions (Table 7) for the twin beds in this study were primarily innerspring (M1, M2, M4, M5, M6, and M7) or solid core foam (M3 and M8) with varying cushioning components, such as containing a FR cotton batting (M1, M2, M4 and M5) or FR-PUF (M3, M6, M7 and M8). Mattresses with vinyl ticking (M2, M6, M7 and M8) were primarily institutional mattresses whereas M1, M3, M4 and M5 were residential mattresses.

In Figure 11, the greatest fire threat is posed by the M7 construction with bedclothes; with PHRR (200 kW) nearly twice the reported values for all other bedding assemblies. Negative values of "normalized" data in Figure 11, which are the bedclothes values subtracted from the combination mattress and bedclothes values, indicates that in most cases the bedclothes alone pose a greater fire threat than the mattress combined with the uniform bedclothes. Only in case of Mattress M7 does the normalized data have positive value suggesting greater fire hazard of the mattress itself. Constructions M2, M3, M4, and M6 may pose the lowest fire threat, as these mattresses generate the lowest reported values for PHRR. Bedding assemblies with these mattresses however yielded more CO (values not reported here) during burning suggesting more incomplete combustion. The fire threat is often assessed based on PHRR and time-to-PHRR (TTP). The latter values were not reported in the study; however, the PHRR for bedclothes alone is greater than most other combinations, which suggests that the fire community may need to consider the impact bedclothes have on fire losses and fire performance criteria in current or proposed mattress regulations.

In 2003, Ohlemiller [19] reported the impact of normal and FR bedclothes on the heat release rates of standard and FR modified mattress sets. Normal bedclothes included filled items consisting of a mattress pad, a comforter and a pillow with polyester (100%) fiberfill, in addition to two sheets, a blanket and a pillowcase. The sets of FR modified bedclothes were of two types: i) mattress pad, comforter and pillows with FR fiberfill and ii) mattress pad, comforter and pillows with FR barriers under their respective cover shells. The FIGRA values derived from Ohlemiller's data are plotted in Figure 12 and Figure 13. Figure 12 compares the impact of mattress pad modifications on standard non-FR as well as FR modified mattresses. In this experiment 4 mattress pads with different fiberfill were used with and without a protective skirt on a standard non-FR PUF mattress and a FR mattress with FR ticking. The details of the mattress pads are given in Table 8. The FIGRA was generally around 10 times higher for the standard PUF mattresses as compared to the FR modified mattresses, except for the bedding assemblies where the mattress pad includes a protective skirt, in which cases the values were more comparable (Figure 12). The protective skirt, which is essentially an extra layer of protection, significantly improves the fire performance of bedding assembly. It is important to note here that fire barrier layers are more effective than FR fillers. Mattress pad A with charring and non charring (thermoplastic) fiberfill in Figure 12 shows the most antagonistic effect on a PUF mattress; that is, it worsens the flammability behavior of the bedding assembly. This may be due to the 'scaffolding effect' of charring and non charring blends, in which the melting thermoplastic envelops the surface of the charring fibers and this developing carbonaceous char prevents any shrinkage of the blended component away from an approaching flame or igniting source [37].

Figure 13 compares the fire threat of FR modified mattresses with a full set of normal and FR modified bedclothes. Detailed description of FR modified mattresses and bedclothes used in this study are given in Table 9. Again, it was noted that the use of a protective skirt further enhances the flame retardance of the bedding assembly in FR modified mattresses.

6. Conclusions

Mattresses vary in size, geometry, construction type, and component materials and are major factors in determining the fire threat of a mattress. The size effect is only significant for standard mattresses without any FR modification. The soft components of a mattress are manufacturer-dependent, with several highly engineered combinations of fibers, fabrics and foams available. All these factors impact flammability of a mattress individually and collectively. In order to allow the mattress manufacturer sufficient flexibility to satisfy the comfort and aesthetic needs of the consumer while still complying with the stringent flammability standards, fire blocking materials appear to be a promising solution.

A fire barrier is part of the overall mattress system. Formulation or physical changes to other components in the system may affect the fire performance of a selected barrier system. As discussed in this report, studies on the impact of fire barriers on the fire performance of mattresses have shown that the performance of fire barriers is strongly dependent on the type of ticking, especially when a flame is used as an ignition source. When used with incompatible combinations of filling materials and tickings, fire barriers may fail to prevent a rise in temperature, smoke and carbon monoxide formation. To date no guidelines exist for the usage of fire barriers in mattress construction. Guidelines for quantifying the performance of fire barriers are also lacking. Currently, the work at NIST is focused on identifying measurement science tools for quantifying the performance of fire barrier materials and for developing materials that may be used to generate a superior fire barrier. Furthermore, the fire hazard from bedclothes cannot be ignored and further research is warranted.

(a)

Number of mattress/bedding-related fires in residential buildings

■ Smoking material ignition
■ Open flame ignition
□ Others

26

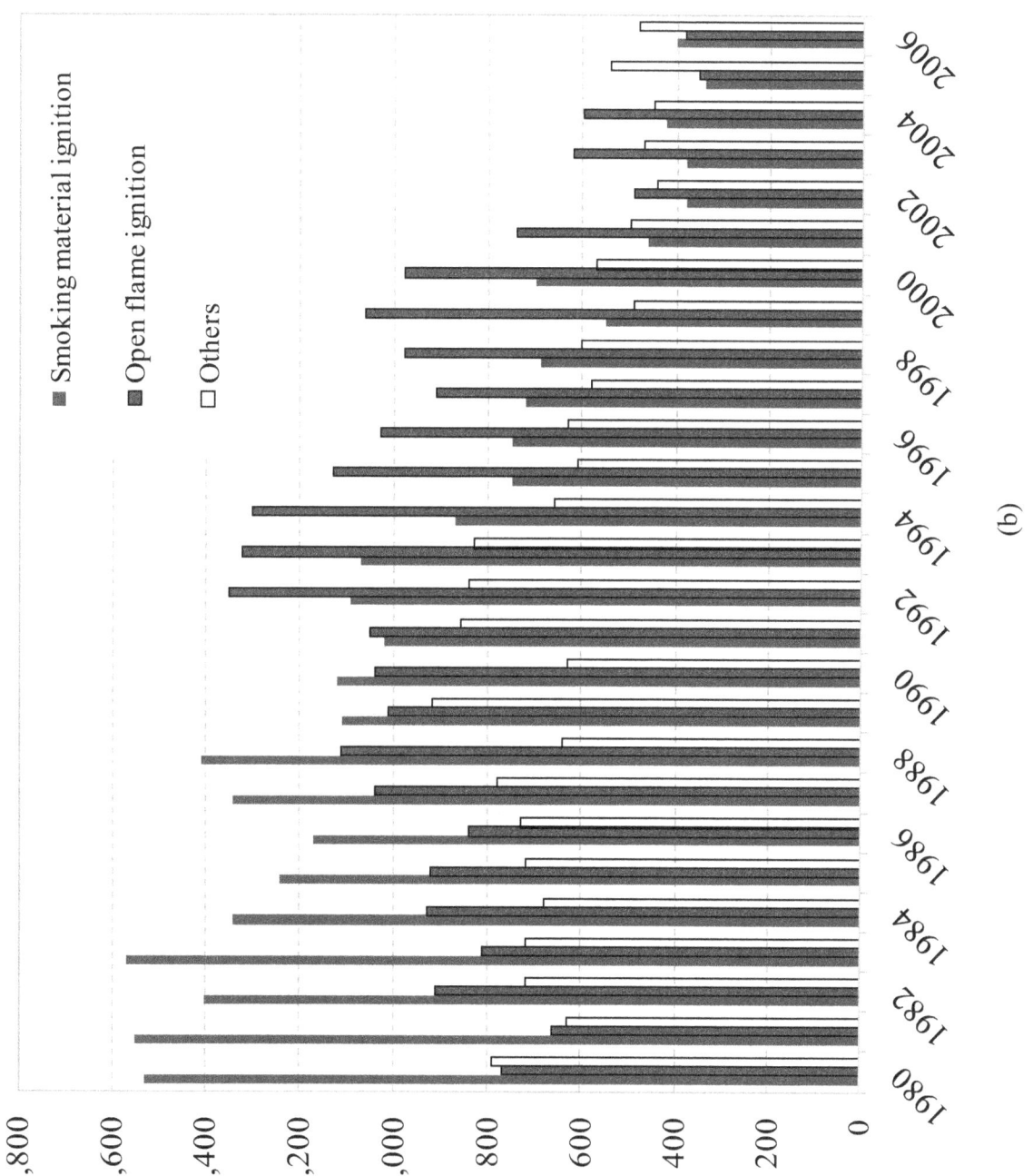

(b)

Number of personal injuries due to mattress/bedding-related fires

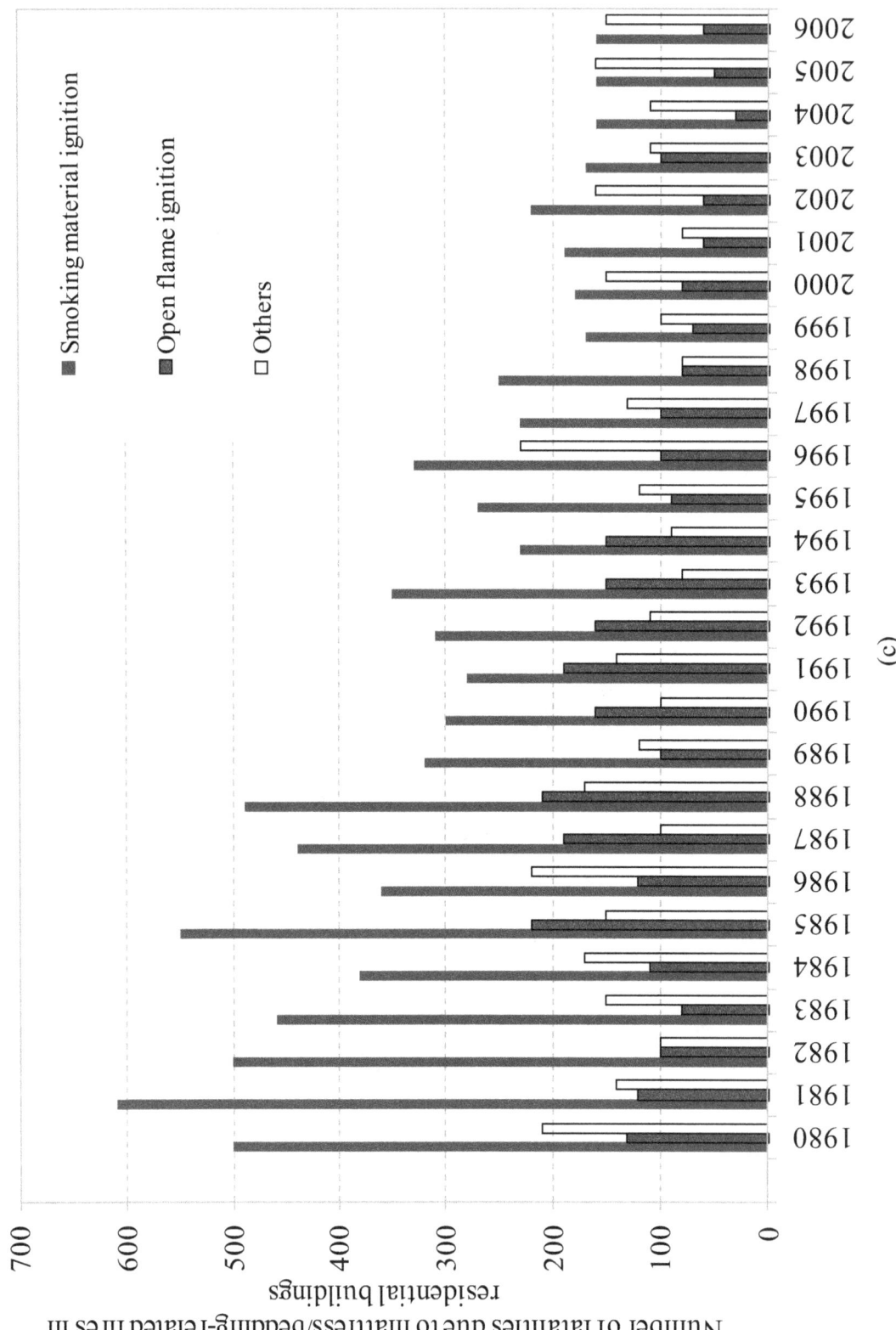

Figure 1. Mattress-bedding related US residential fire losses from 1980 to 2006, delineated by ignition source. Number of (a) fires, (b) civilian injuries, and (c) civilian fatalities [13].

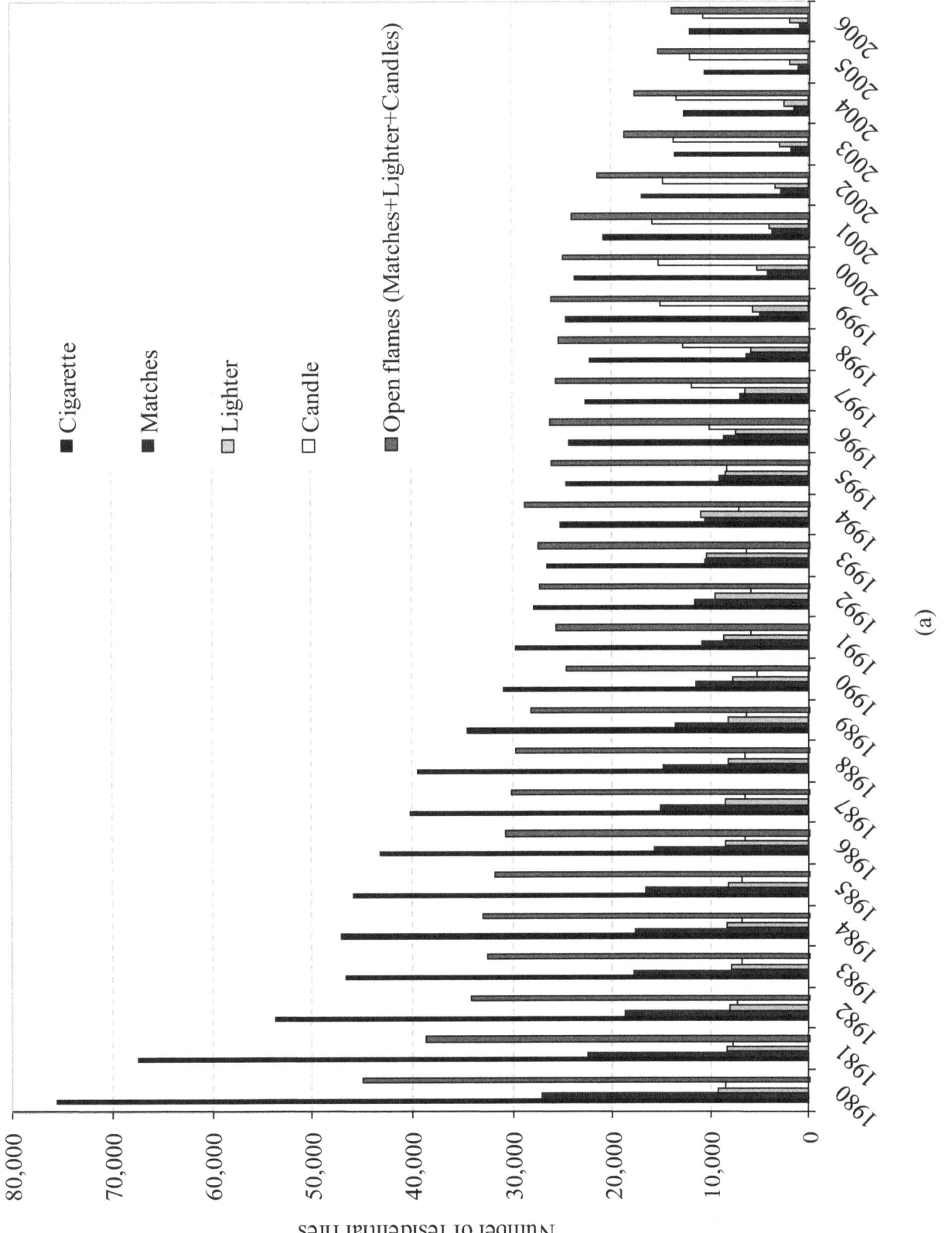

(a)

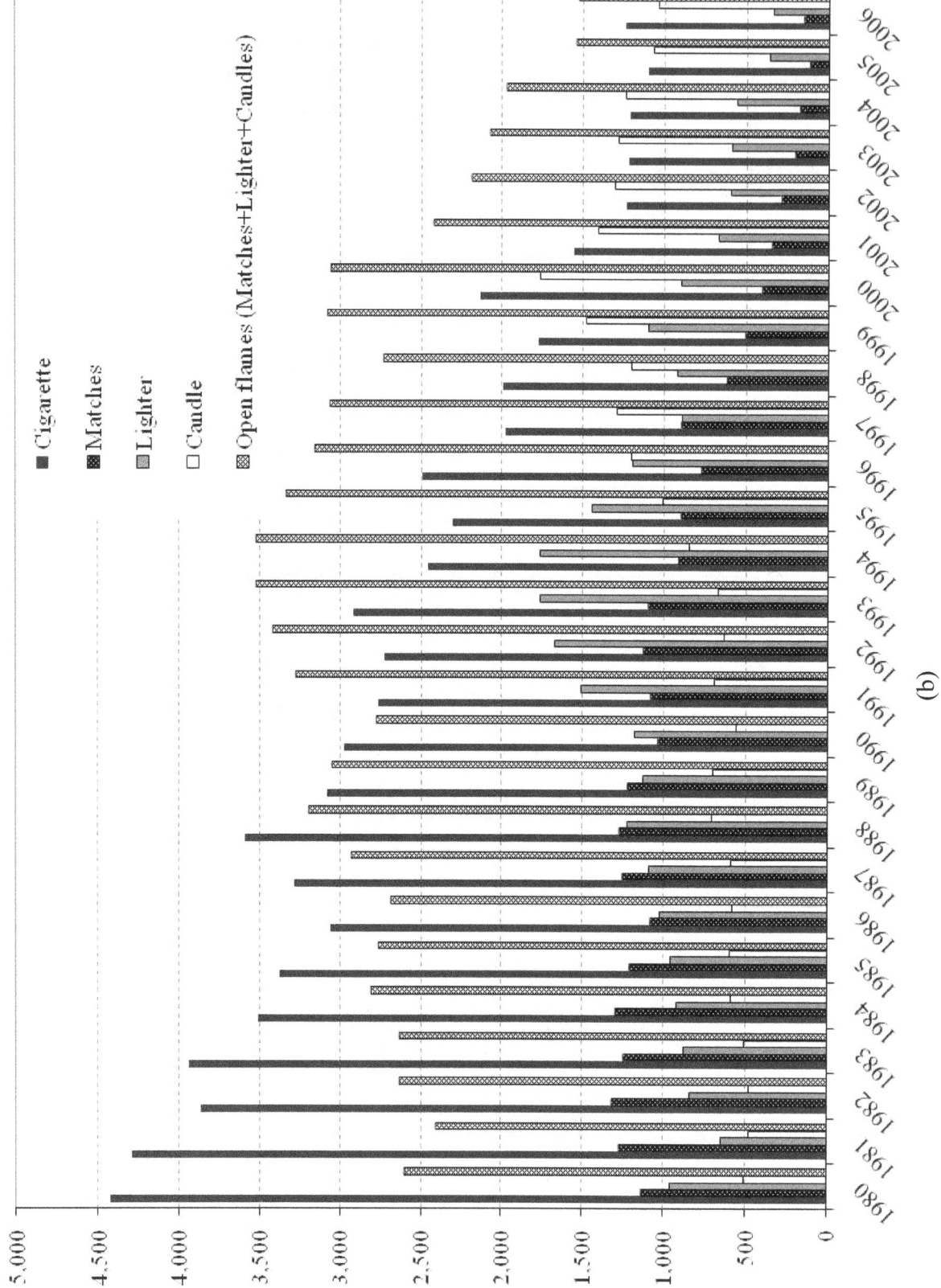

(b)

Number of personal injuries in residential fires

30

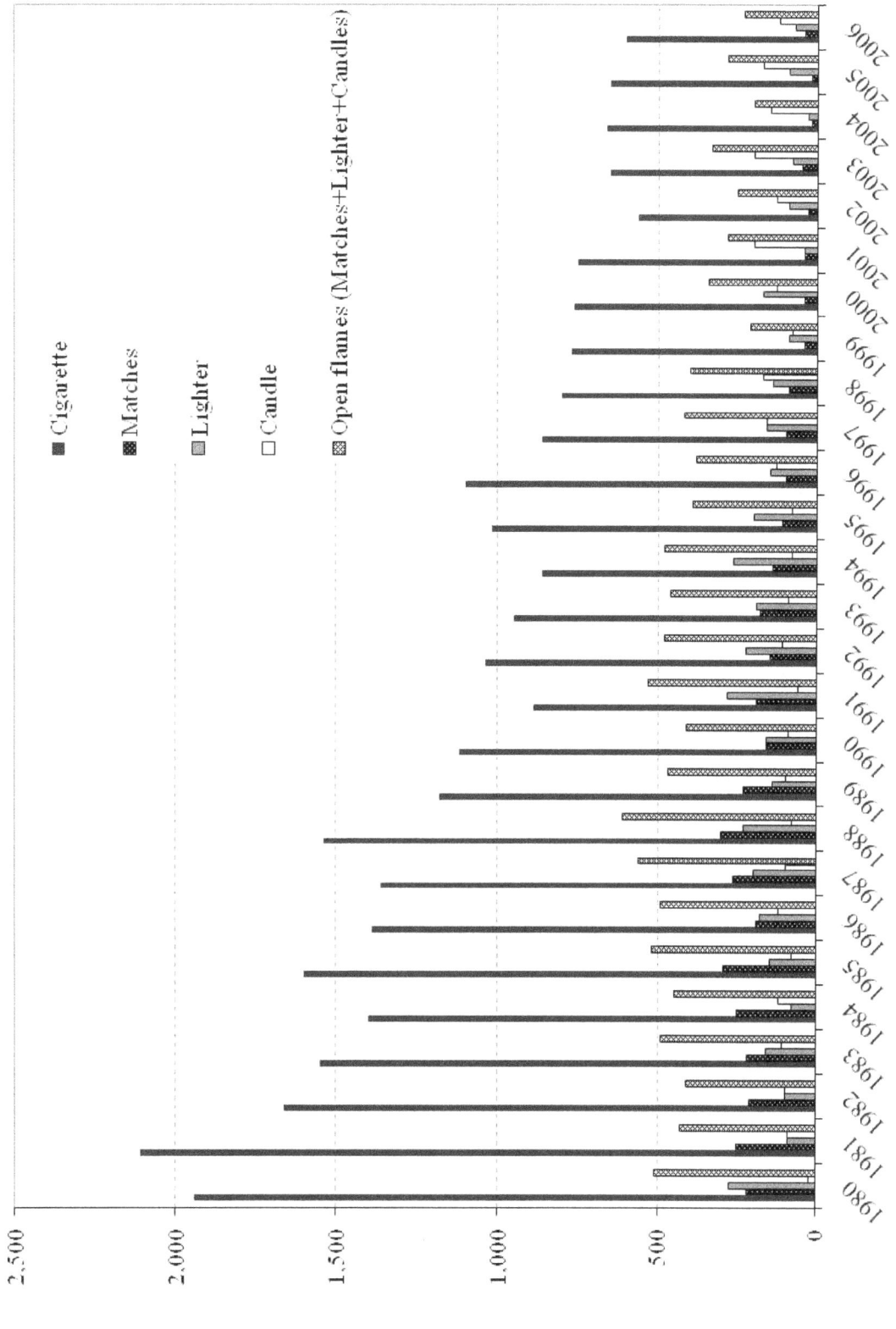

(c)

Figure 2. 1980 to 2006 US residential fire losses according to ignition source. Number of (a) fires, (b) civilian injuries, and (c) civilian fatalities [13].

31

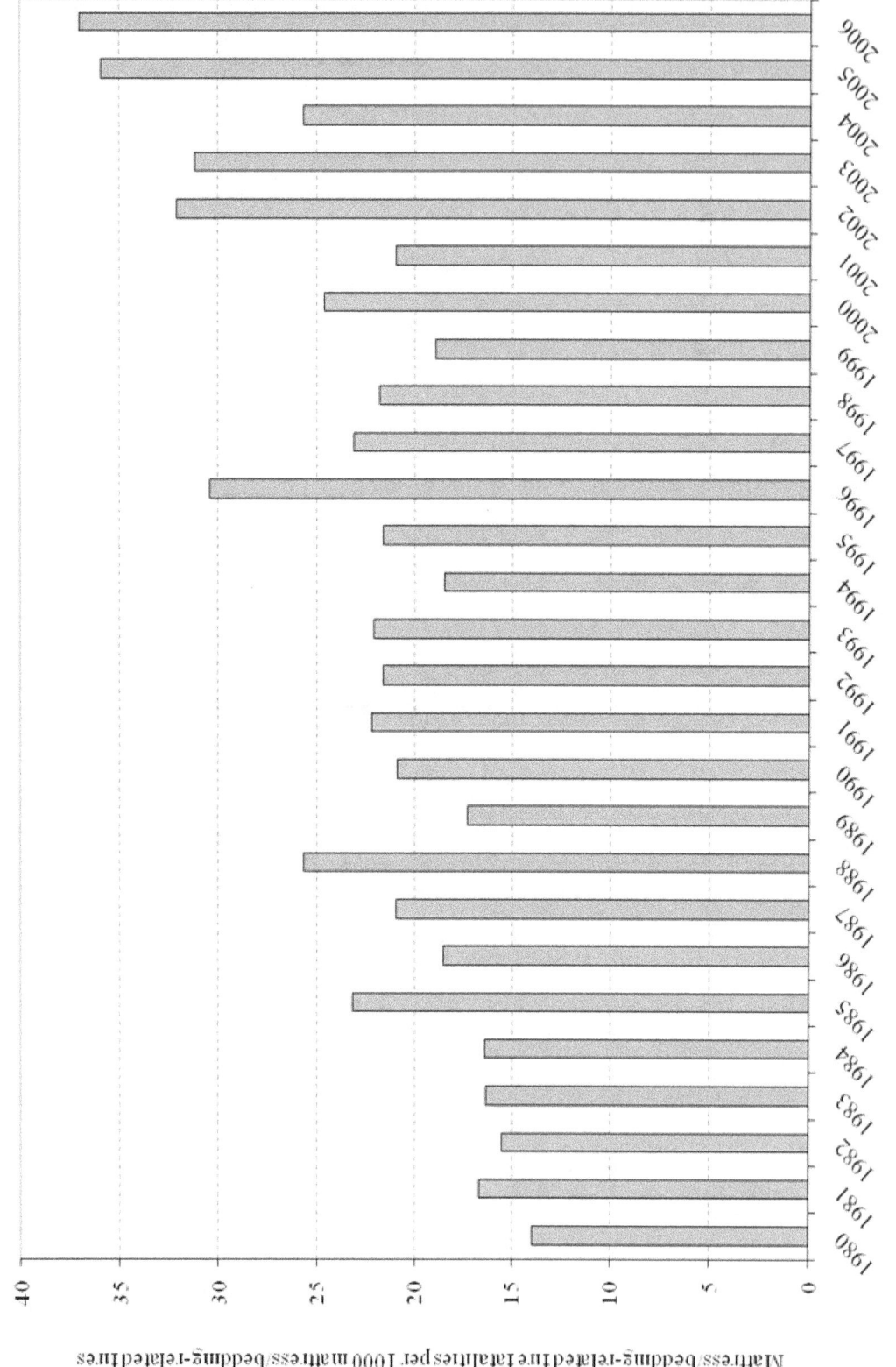

Figure 3. Mattress/bedding-related US residential fire losses from 1980 to 2006; civilian fatalities per 1000 mattress/bedding fires (for all ignition sources) [13].

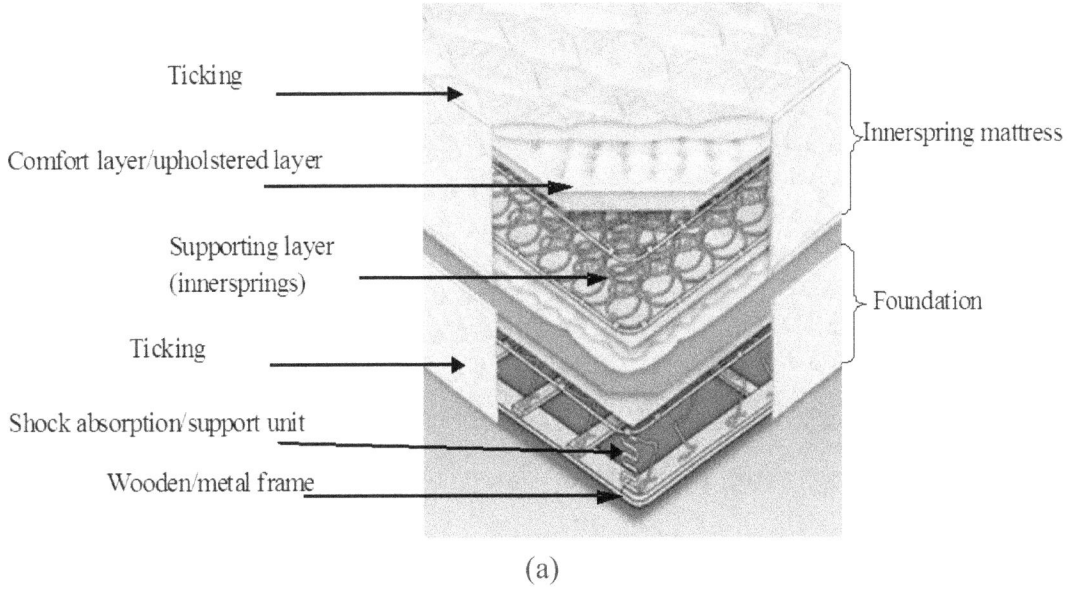

(a)

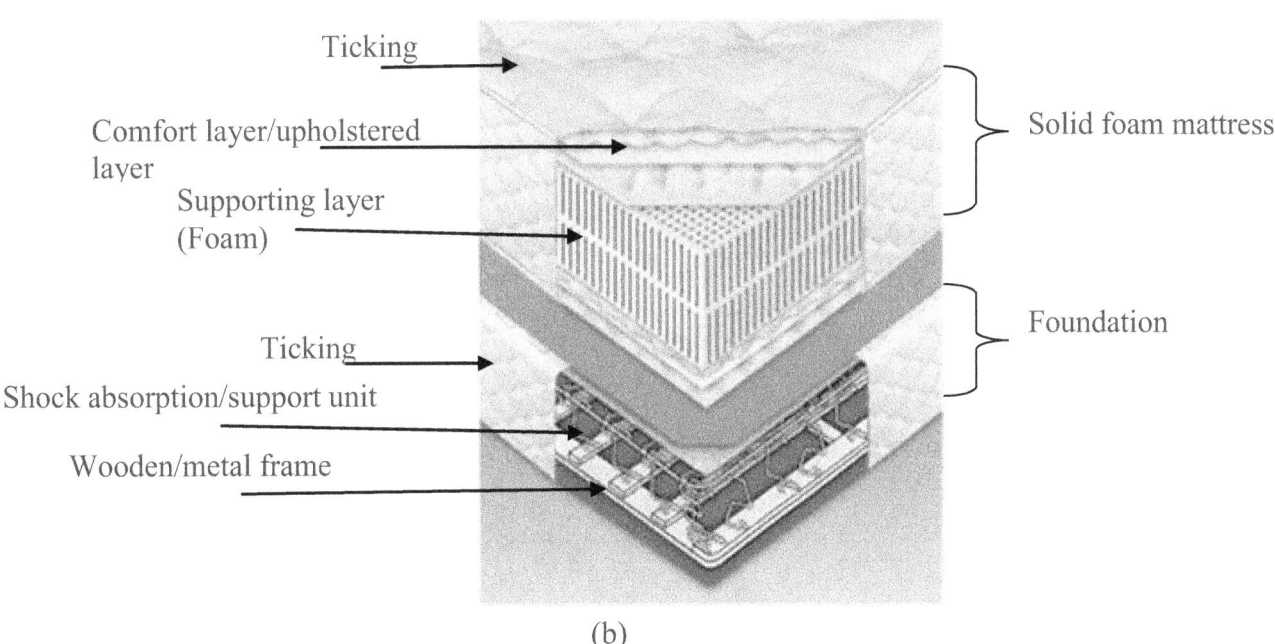

(b)

Figure 4. Schematic of a typical residential mattress and foundation set: (a) innerspring and (b) solid foam [38].

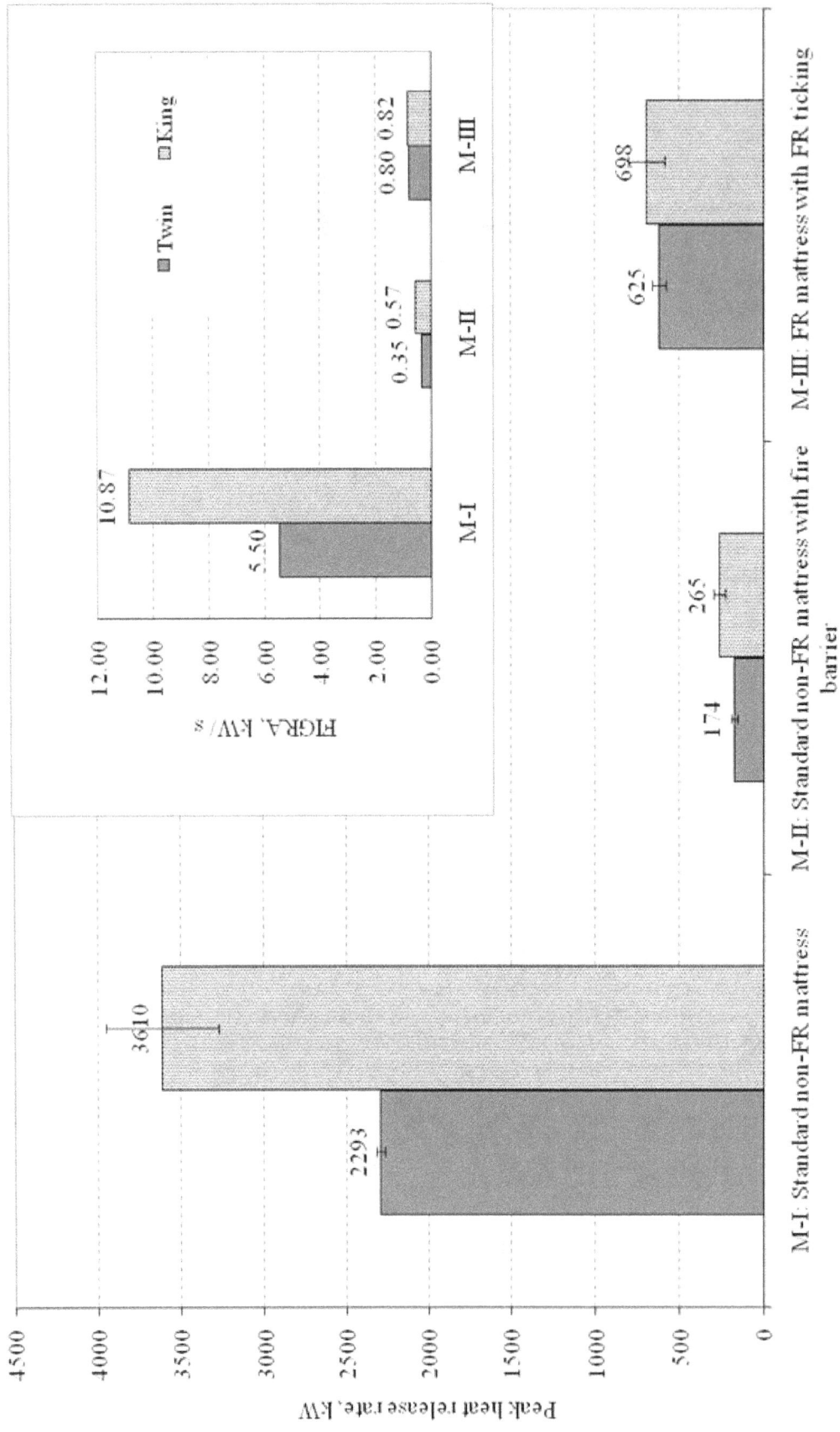

Figure 5. Impact of bed sizes on fire performance of (i) M-I Standard non-FR mattress, (ii) M-II Standard non-FR mattress with fire barrier and (iii) M-III FR mattress with FR ticking [6, 19]. Error bars represent relative standard deviation of average PHRR values (average of two replicates).

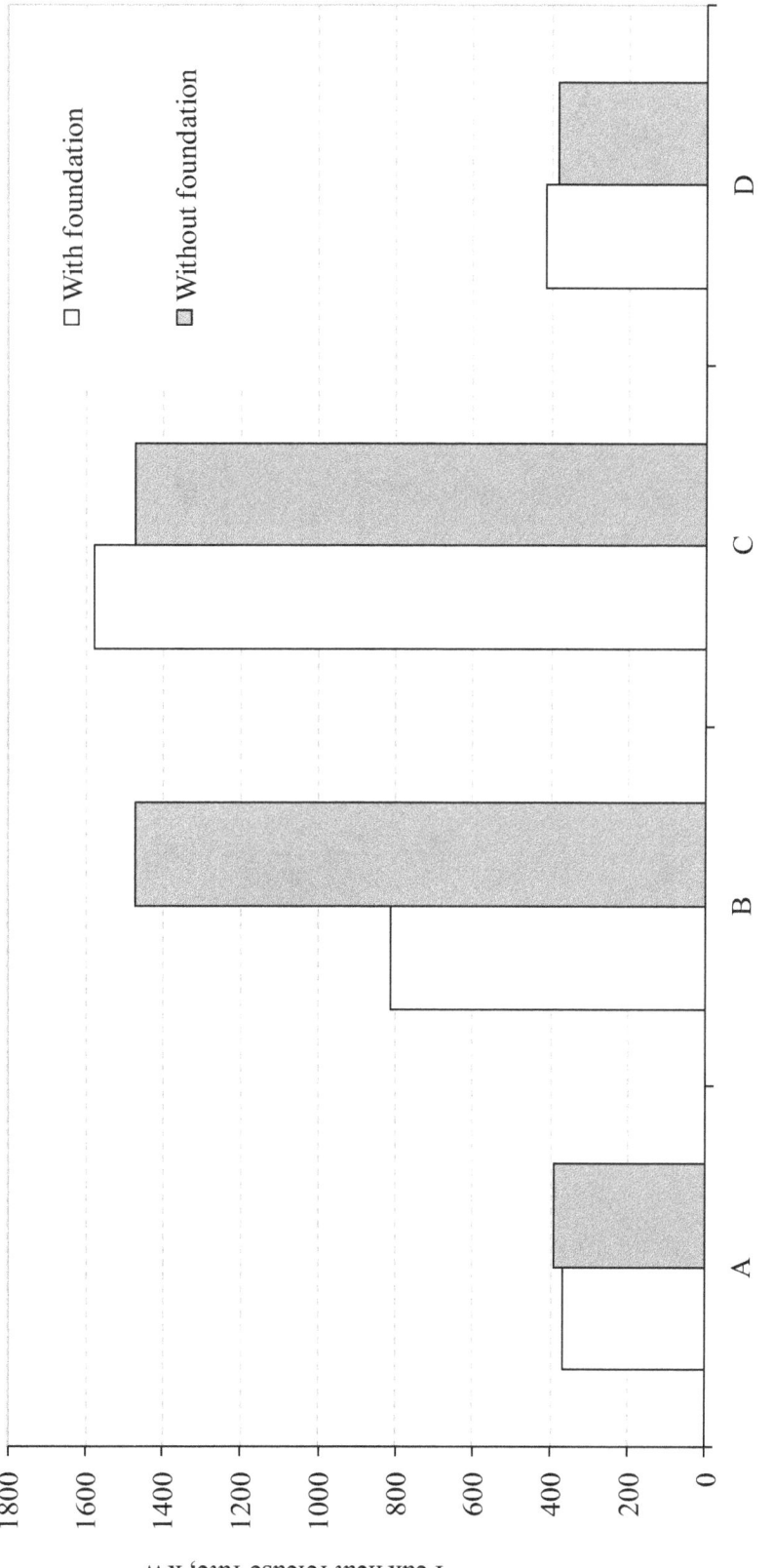

A: Innerspring mattresses with conventional non-FR polyurethane foam filling.
B and C: Similar mattress construction with box spring foundations containing cotton batting and foam fillings respectively.
D: Innerspring mattresses with California TB 117 grade foam.

Figure 6. Impact of foundation on fire performance of mattresses [29].

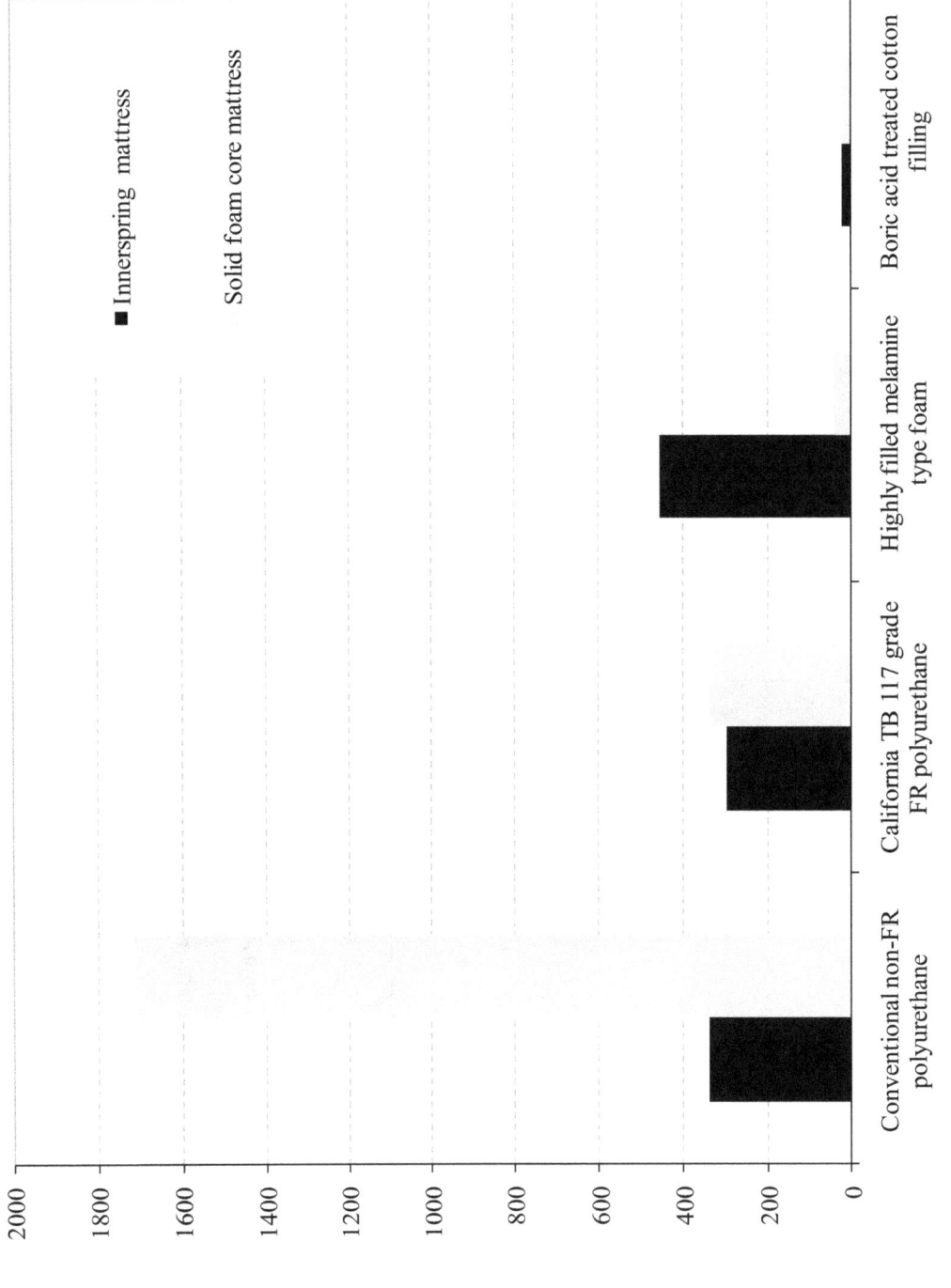

Figure 7. Impact of mattress construction on peak heat release rate: Innerspring vs solid core [29].

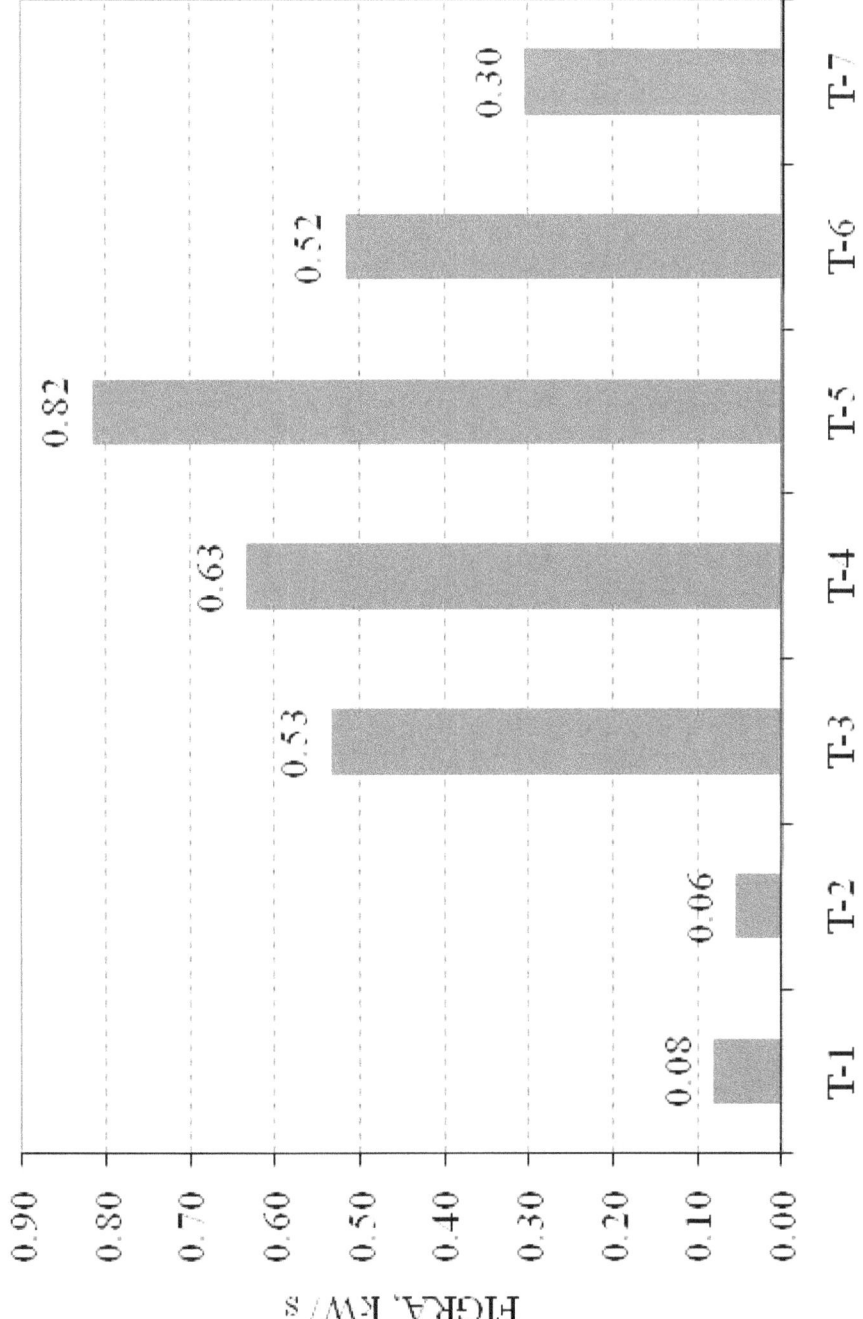

Figure 8. Impact of ticking on fire performance of mattresses [33].

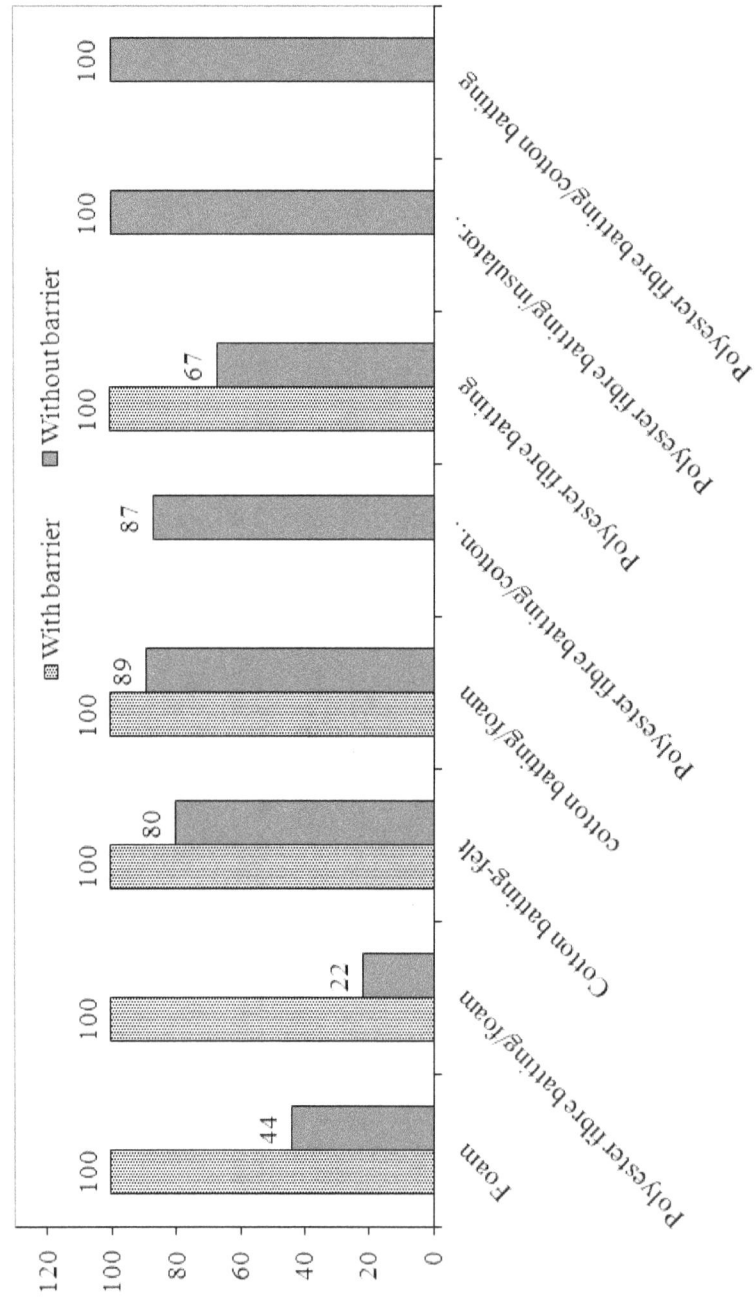

Figure 9. Comparison of innerspring mattresses with different type of filling materials in presence or absence of barrier fabrics [16].

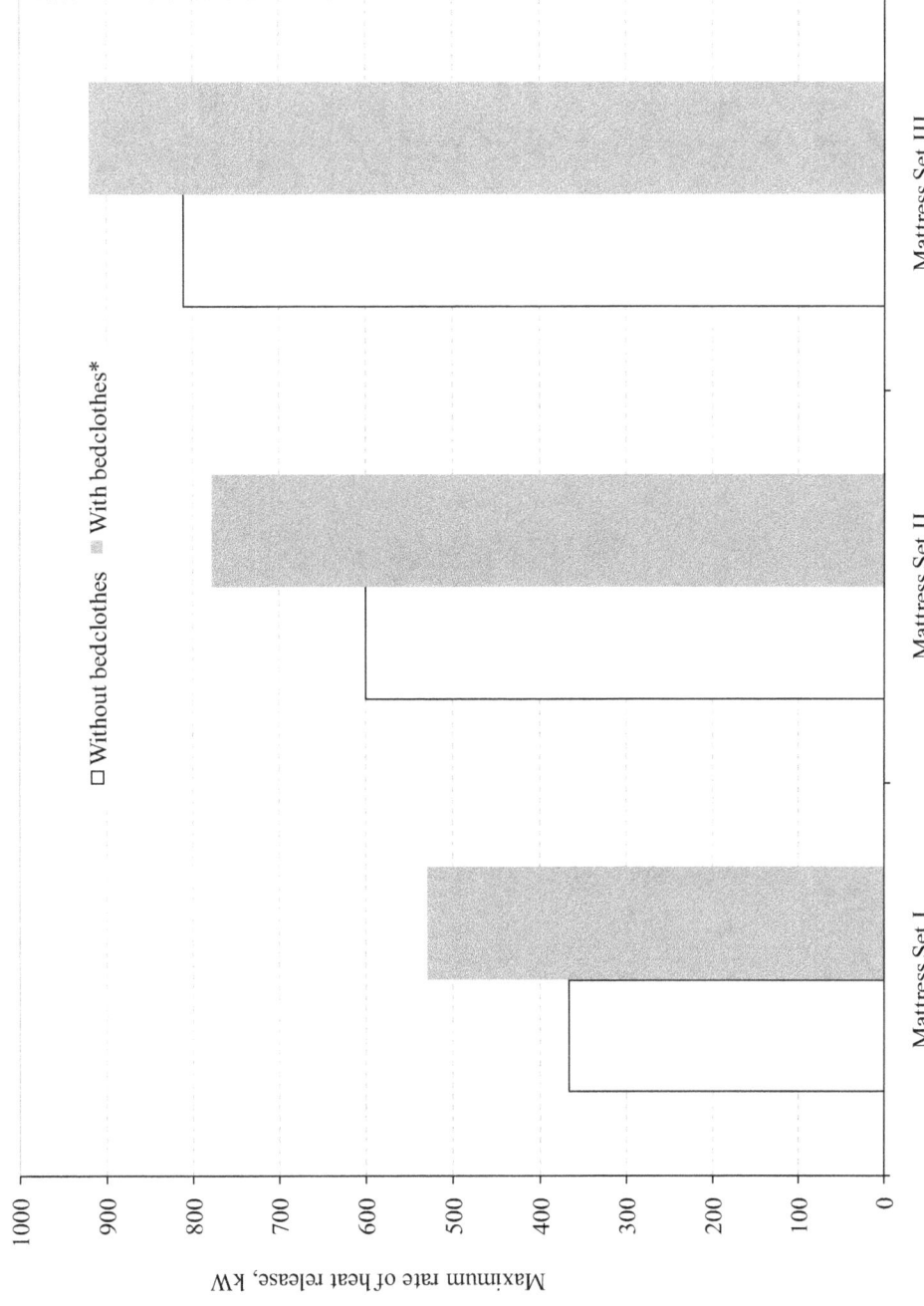

Figure 10. Impact of bedclothes on fire performance of various mattress/foundation sets. [29] *Bedclothes include a mattress pad, two bed sheets, pillow with a pillowcase, and two blankets.

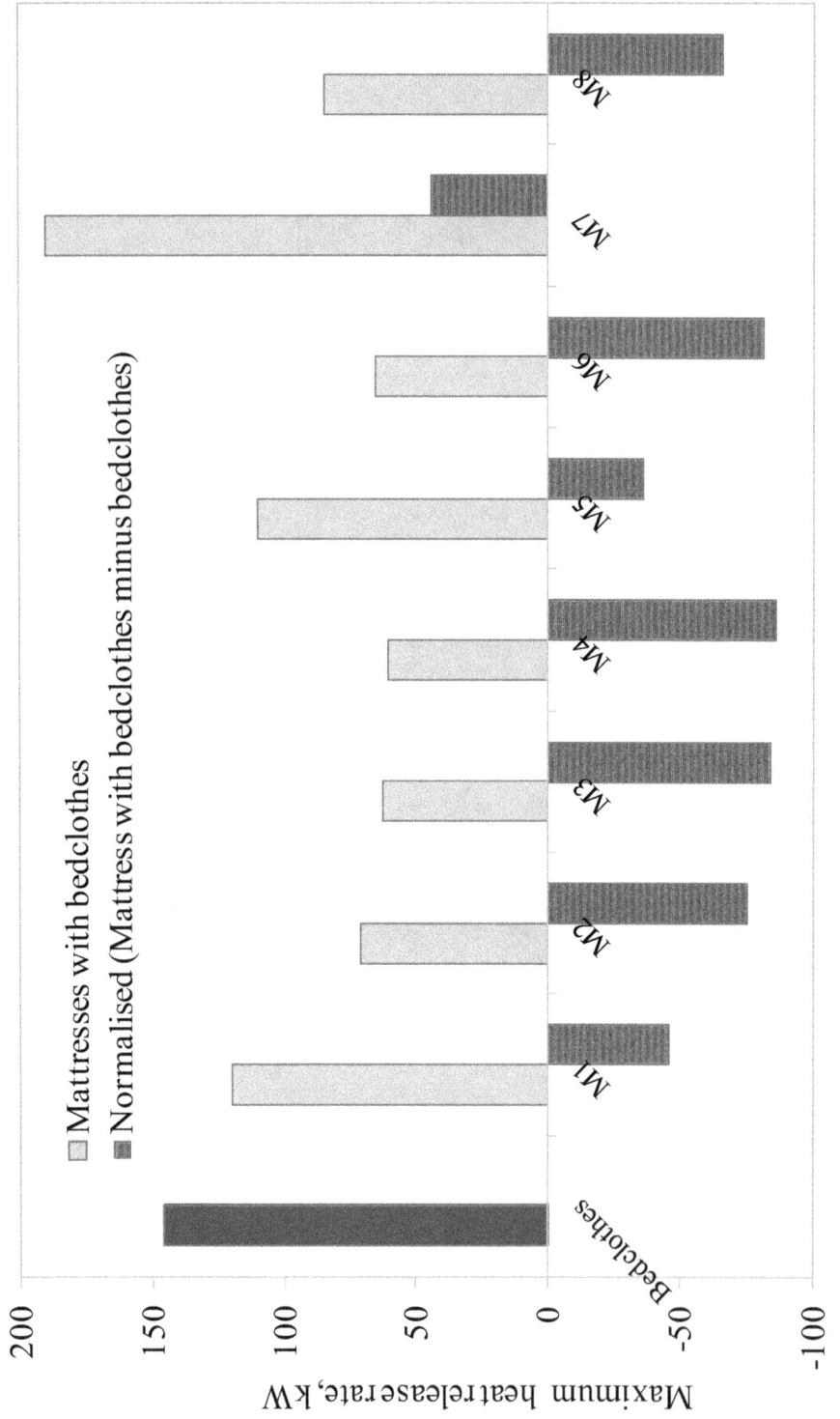

Figure 11. Impact of normal bedclothes on fire performance of mattresses having varying levels of fire retardance [29].

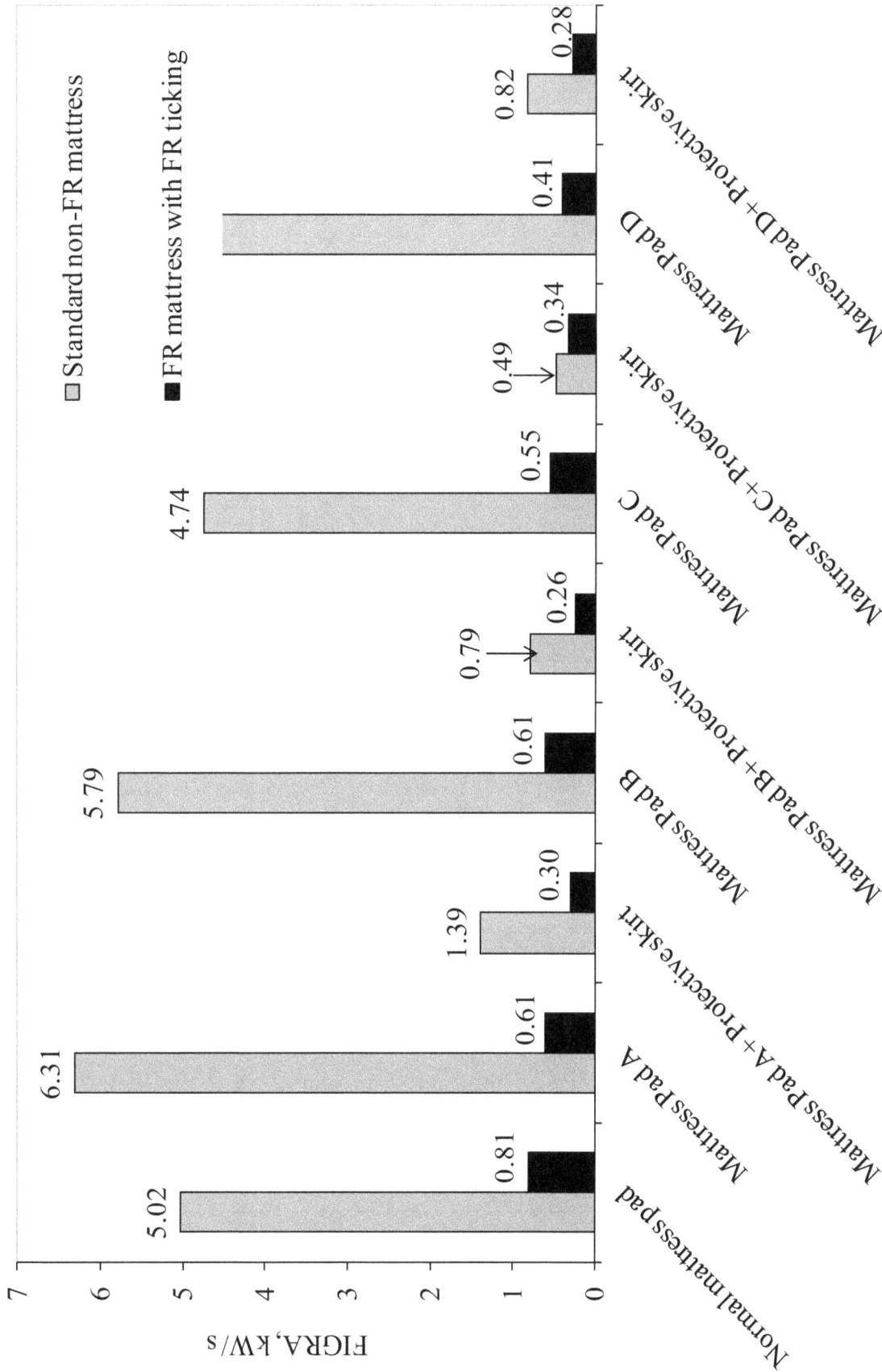

Figure 12. Impact of mattress pads on fire performance of FR and non-FR mattresses [19].

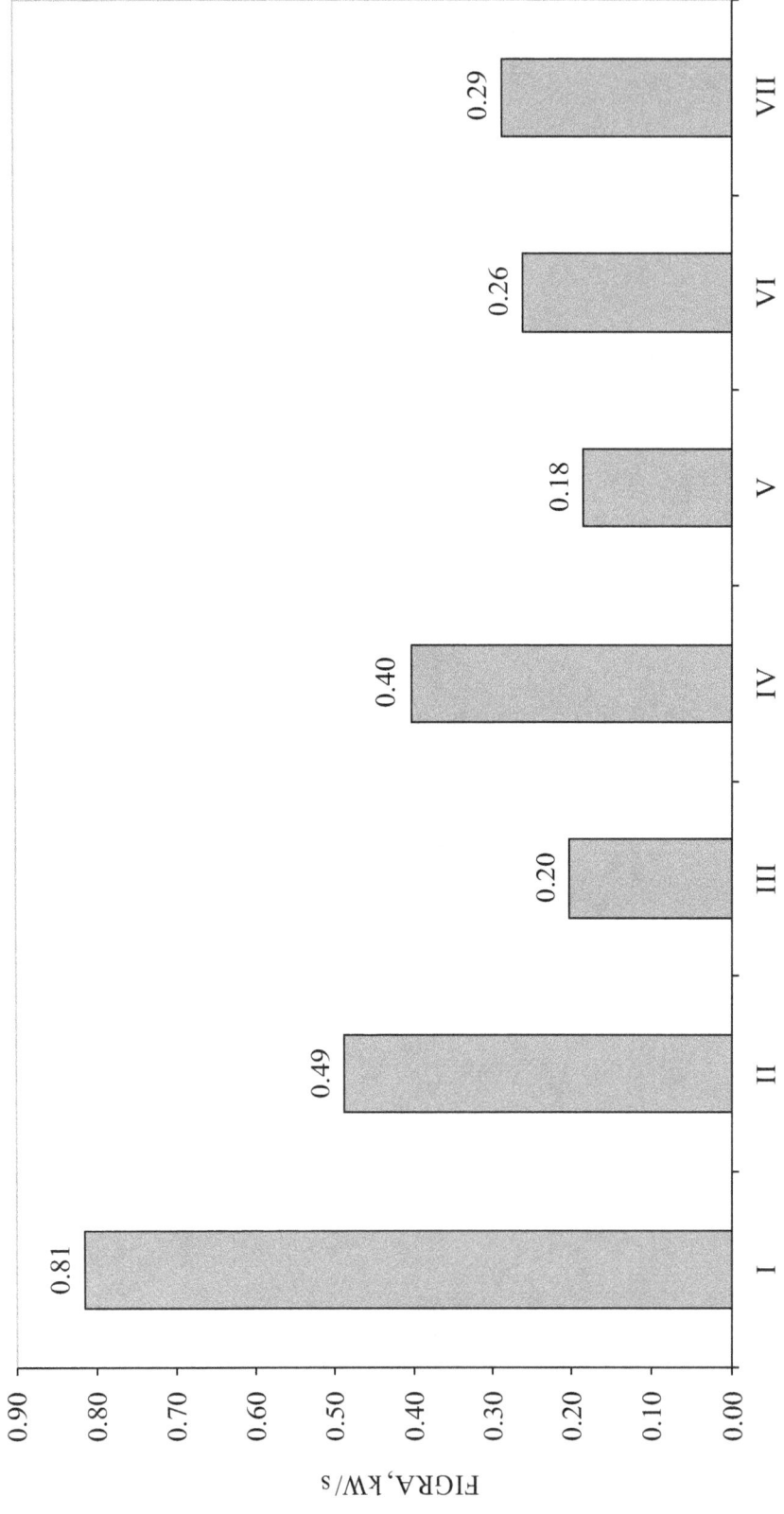

Figure 13. Impact of bedclothes on fire performance of FR modified mattresses [19].

Table 1. Standards and test methods for mattress and beddings

Issuing Authority/Country	Standard Code (Effective/Revised date)	Scope	Measured parameters
United States			
Consumer Product Safety Commission (CPSC)	16 CFR 1632	Prescribes a test procedure for determination of ignition resistance of residential mattress when exposed to lighted cigarette.	Char length in any direction from the nearest point of the cigarette.
	16 CFR 1633 (Effective July 1, 2007)	Standard for the flammability (open flame) of residential mattress sets	Peak and total heat release rate.
BHFTI (State of California, Department of Consumer Affairs)	Cal TB 129	Flammability test procedure for mattresses for use in public buildings	Peak and total heat release rate, mass loss in open calorimetry test.
	Cal TB 603 (January 2005) (now superseded by 16 CFR 1633)	Test procedure for open-flame fire testing of residential mattresses under well-ventilated conditions.	Rate of heat release in oxygen consumption calorimetry and burning behavior.
	Cal TB 604: (January 2005, Rule making suspended in March 29, 2010) Section 1 Section 2 Section 3	Flammability (open flame) standard for filled bedclothes: Comforters and bed spreads Pillows and bed cushions Mattress pads	Percentage weight loss Percentage weight loss Burning behavior
	Cal TB 106 Superseded by 16 CFR 1632	Resistance of mattress or mattress pad to cigarette ignition.	Char length in any direction from the nearest point of the cigarette.
	Cal TB 121	Flammability of mattresses used in high risk occupancies subjected to a galvanized metal container with ten (10) double sheets of loosely wadded news paper	Mass loss, change in temperature at the ceiling and CO production
Boston Fire Department (Boston,	Boston Fire Department Method IX-11	Mattresses (with bedclothes) intended for use in health care facilities, hotels and dormitories	Full scale burning behavior using furniture calorimeter

Organization	Standard	Description	Details/Criteria
Massachusetts)	Michigan Roll-Up Test	For mattresses used in jails	Mattress or pads are rolled up, tied and stuffed with newspaper and leaned against the bed frame. There is no specified test criteria
American Standard Test Methods (ASTM)	ASTM E-1590	Standard test method for determination of burning behavior of mattresses used in public occupancies	Rate of heat release by an oxygen consumption method, production of light-obscuring smoke and the concentrations of certain toxic gas species in the combustion gases
	ASTM D 7140	Test method to measure heat transfer through textile thermal barrier materials.	Heat transfer properties of barrier material when exposed to a calibrated convective and radiant energy heat source for 60 seconds
	ASTM D 7016	Test method to evaluate edge binding components (e.g. thread, tape)used in mattresses after exposure to an open flame	Flammability characteristics of mattress edge bindings and sewing threads during and after exposure to an open flame ignition source.
National Fire Protection Association (NFPA)	NFPA 267	Standard method of test for fire characteristics of mattresses and bedding assemblies exposed to flaming ignition source	Heat release, smoke density, weight loss, and generation of carbon monoxide of mattresses and bedding assemblies using an open calorimeter environment.
International			
Canada			
Underwriters' Laboratories (UL)	CAN/ULC-S137	Standard test method for fire growth of mattresses (open flame test)	Measures PHRR, THR and mass loss when subjected to a specified flaming ignition source under well ventilated conditions
	UL 1895	Fire tests of mattresses	Investigates the ability of a mattress to resist rapid heat release when subjected to a flaming ignition source.
	UL 2060 (withdrawn)	Standard for fire test of mattresses with bedclothes using a furniture calorimeter	Investigates the ability of a mattress to resist rapid heat release when subjected to a flaming ignition source.
United Kingdom			
British Standards Institution (BS)	BS EN 597: 1995 (Replaced BS 6807:1990)	Assessment of the ignitability of mattress sets Ignition source: smoldering cigarette. Match flame equivalent	Burning behavior: Unsafe escalating combustion Smoldering through thickness Char length

			Flaming ignition in case of match-flame equivalent ignition source.
	BS 7177:2008	Specification for resistance to ignition of mattresses, mattress pads, divans and bed bases	-
	BS 7175:1989	Methods of test for the ignitability of bedcovers and pillows by smoldering and flaming ignition sources	Burning behavior observed for : Hole formation, melting, dripping, charring, ignition and development of flames from smoldering.
Sweden			
Swedish Standards Institute/Sweden	SS EN 597:1994	Same as BS EN 597 : 1995	
	SS 876 00 10	Hospital beds, high performance	
Denmark			
Denmark	NT FIRE 037	Procedure to determine the ignitability of bedclothes , including mattress pad with small smoldering and flaming sources of ignition.	Individual component test
Germany			
German Institute of Standards (DIN)	DIN EN 14533	Textiles and textile products - Burning behavior of bedding items - Classification scheme	-
Others			
International Maritime Organization (IMO)	IMO MSC. 61(67), Annex 1, Part 9, MO Res A.688 (17)	Ignitability of bedding components	As mentioned in NFPA 267, ASTM 1590 16 CFR 1633
ISO	ISO 12952-2:1998	Burning behavior of bedding items -- Part 2: Specific test methods for the ignitability by a smoldering cigarette	Char length, smoldering and formation of holes

Table 2. Scope of variations in mattress designs

Mattress parameters	Description	Variations
Height	Between 10.16 cm and 50.8 cm (4 in and 20 in)	
Sizes	Twin (96.5 cm x 187.9 cm (38 in x 74 in)), Full (134.6 cm x 187.9 cm (53in x 74 in)), Queen (152.4 cm x 203.2 cm (60 in x 80 in)), King (193.0 cm x 203.2 cm (76 in x 80 in)) and California King (182.8 cm x 213.3 cm (72 in x 84 in))	5
Construction	Single-sided and double-sided	2
Mattress geometry	Smooth top, Tight top, Pillow top, Super pillow top, Euro top, Box pillow top.	6
Mattress core	Open coil with or without foam encasement, Pocket coil with or without foam encasement, Foam, Viscoelastic, Latex, and Air	8
Foundation geometry	Steel/wood box continental (22.8 cm (9 in)) and 7.6 cm (3in)); Steel/wood Taped (22.8 cm (9 in) and 7.6 cm (3in)); Wood box (22.8 cm (9in)); Wood box, cardboard taped (5.0 cm (2in)); No box	7
Upholstery/ fillings	Numerous combinations of fiber, fabric and foams	100
Ticking	Highly variable	> 1000
Total variations excluding ticking variations: 5 x 2 x 6 x 8 x 7 x100 =		**> 336,000**

Table 3. Heat release test data for various mattress constructions [29]

Mattress description			PHRR (kW)	Ceiling temperature (°C)	Mass loss in 10 min (kg)
Type	Filling	Other			
Innerspring	Conventional non-FR PUF	Without foundation	337 ± 57	376 ± 104	8.300
		With foundation	366	375.5	9.389
		With bedclothes*	528	454	11.249
	California TB 117 grade FR PUF foam	Without foundation	296 ± 84	277 ± 61	7.574
		With foundation	416	400	9.480
	Highly filled melamine type foam	-	453 ± 95	458 ± 59	9.842
	Neoprene type foam	-	48 ± 28	75.5	0.589
	Neoprene type foam	-	50 ± 24	100.5	0.589
	2.54 cm (1 in) conventional non-FR PUF pad / Shredded polyester fiber insulator pad	Reinforced vinyl cover	335	282	4.762
	1.27 cm (½ in) conventional non-FR PUF pad / FR cotton batting-boric acid treated / Shredded polyester fiber insulator pad	Reinforced vinyl cover	29	75.5	0.181
	FR cotton batting-boric acid treated / Boric acid treated insulator pad	Reinforced vinyl ticking	17	69.4	0.090
	FR PUF foam	Vinyl ticking, Topper pad of FR foam and glass barrier	65	123	1.723
	FR cotton batting	Woven ticking with aluminized barrier	100	142	2.857

	FR cotton batting	Vinyl ticking	71	127	2.086
	FR cotton batting/ FR insulator pad	Non-woven FR barrier quilted to Cal 117 foam and FR woven ticking,	60	129	2.313
	Fire barrier (thin layer of highly fire resistant cellular foam bonded to fiberglass fabric)	-	22	63.3	0.226
	Fire barrier (highly engineered fire resistant cover fabric)	-	20	70	0.090
	Boric acid treated cotton filling	-	22 ± 1	65.5	0.408 ± 0.206
Solid core foam	FR PUF foam	Vinyl ticking	85	117	2.117
	Highly filled melamine type foam	Woven ticking	62	134	1.587
	Boric acid treated cotton	-	22	73.8	0.362
	Conventional non-FR PUF	-	1716	1031	3.764
	California TB 117 grade FR PUF	-	339	406	7.983
	Highly filled melamine type foam	-	39	86.6	0.816

* Bedclothes include a mattress pad, two bed sheets, a bed pillow with pillowcase and two blankets.

Table 4. Heat release data for residential mattresses [32].

Mattress construction	Combustible mass (kg)	PHRR (kW)	FIGRA (kW/s)	Effective heat of combustion (MJ/kg)	THR (MJ)	Total smoke released (m^2)
Innerspring mattress with thick PUF pillow top -02	9.2	2038	8.78	24.9	232	42
Innerspring mattress with thin PUF pillow top -03	5.3	1648	9.75	24.3	131	29
Innerspring mattress with foam encased pocket coils-06	10.5	3496	15.33	24.5	256	126
Solid core with three layers of PUF-01	12.5	3493	15.25	18.5	231	94
Solid core mattress with viscoelastic foam top-04	13.3	3433	12.57	22.5	300	132

Table 5. Classification of ticking according to 1632.6 of 16 CFR 1632 [1].

	Test procedure	Performance requirements	Ticking characteristics	Substitution procedure
Class A	Three ticking prototypes tested directly over cotton batting on the test boxes and Three ticking prototypes tested over urethane foam covering the cotton batting on the test boxes	All six specimens meet the test criteria (char length < 2.54 cm (1 in), cotton batting does not ignite).	Acts as barrier against cigarette ignition.	May be used on any qualified mattress prototype without conducting new prototype test.
Class B	Three ticking prototypes tested over PUF covering the cotton batting on the test boxes	The three specimens tested over PUF meet the test criteria. One or more specimens tested over cotton batting do not meet the test criteria.	Has no effect on cigarette ignition.	May be used on any mattress prototype which was qualified with Class B or Class C without conducting new prototype tests.
Class C	Three ticking prototypes tested over urethane foam covering the cotton batting on the test boxes	One or more specimens tested over PUF covering cotton batting do not meet the test criteria.	Has the potential to act as a contributor to cigarette ignition.	Requires a new mattress prototype test before the ticking fabric is used in mattress production.

Table 6. Heat release test data for mattresses with different tickings [33].

Tick ID	Fabric content (mole fraction %)	Finish	PHRR (kW)	TTP* (s)	THR in first 10 mins (MJ)	FIGRA** (kW/s)
T-1	Polypropylene (100)	Pigment print	73	900	12.7	0.08
T-2	Polypropylene/Polyester (50/50)	Heat set softener	76	1380	13.3	0.05
T-3	Polypropylene/Polyester (32/68)	Hot melt	32	60	3.9	0.53
T-4	Polyester (100)	Aqueous scour	38	60	8.1	0.63
T-5	Polyester (100)	Latex	49	60	14.7	0.81
T-6	Cotton/Polyester (75/25)	Bleach	31	60	5.4	0.51
T-7	Rayon/Polyester (54/46)	none	31	102	6.8	0.30

* TTP: Time to peak heat release rate, ** FIGRA: Fire growth rate index.

Table 7. Details of mattress construction and components [29].

Mattress ID	Construction type	Filling	Ticking	Application
M1	Innerspring	FR cotton batting	Woven ticking with aluminized barrier	Residential
M2	Innerspring	FR cotton batting	Vinyl	Institutional
M3	Solid foam	FR polyurethane (melamine foam)	Woven fabric	Residential
M4	Innerspring	FR cotton batting+ insulator pad	FR Woven fabric, Non-woven barrier quilted to Cal 117 foam and ticking	Residential
M5	Innerspring	FR cotton batting+ insulator pad	Woven ticking quilted to Cal 117 foam	Residential
M6	Innerspring	FR polyurethane foam/Topper pad of FR foam and glass barrier	Vinyl	Institutional
M7	Innerspring	FR polyurethane CMHR foam	Vinyl	Institutional
M8	Solid foam	FR polyurethane foam	Vinyl	Institutional

Table 8. Description of mattress pads and fiberfill [19].

Mattress pad description	Top shell	Fiberfill	Bottom shell
Normal mattress pad	100 % cotton	100% polyester	Non-woven scrim
Mattress pad A		Blend of charring/non-charring fibers	
Mattress pad B		Charring fiber	
Mattress pad C		100% polyester charring barrier fabric	

Table 9. Description of FR modified bedding assemblies and FR bedclothes [19].

Bedding assembly*	Protective skirt	Description of bedclothes				
		Mattress pad	Comforter	Pillow	Fitted, flat sheets, and pillowcase	Blanket
I	X	100 % polyester fiberfill and100 % cotton shell			50 % cotton/50 % polyester	100 % acrylic
II	X	Blend of charring and non-charring fiberfill	FR polyester fill			
III	√					
IV	√	100 % polyester fiberfill and charring barrier fabric (type 1) under 100 % cotton shell				
V	√					
VI		100 % polyester fiberfill and charring barrier fabric (type 2) under 100 % cotton shell				
VII	√					

* Bedding assembly includes mattress set (mattress and foundation) and bedclothes including protective skirt, mattress pad, two bed sheets, pillow with a pillowcase, blanket, and a comforter.

7. References

[1]   16 CFR 1632 Standard for the flammability of mattresses and mattress pads. Consumer Product Safety Commission. May (1991). Available from: http://www.cpsc.gov/businfo/testmatt.pdf.

[2]   Tohamy SM, Aiken DV. Assessing regulatory effectiveness with exogenously declining risk: A case study of the CPSC's 1973 mattress standard, Journal of Safety Research. 2007; 38: 661-668.

[3]   Ahrens M. Home fires that began with upholstered furniture. National Fire Protection Association. May (2008). Summary available from: http://www.nfpa.org/assets/files/PDF/UpholsteredExecutiveSum.pdf

[4]   Greene M, Miller D. 2006-2008 Residential Fire Loss Estimates. Consumer Product Safety Commission report. August (2010). Available from: http://www.cpsc.gov/library/fire06.pdf.

[5]   Fleischmann CM. Flammability tests for upholstered furniture and mattresses, Chapter 7 in *Flammability Testing of Materials used in Construction, Transport and Mining*, Apte VB, Editor, Woodhead Publishing, Cambridge, UK, 164-186, (2006).

[6]   Ohlemiller TJ and Gann RG, Estimating reduced fire risk resulting from an improved mattress flammability standard, NIST Technical Note 1446, August 2002. National Institute of Standards and Technology, Gaithersburg MD.

[7]   BS 6807: Methods of test for assessment of ignitability of mattresses, upholstered divans and upholstered bed bases with flaming types of primary and secondary sources of ignition, British Standards Institution, London.

[8]   BS EN 597: Furniture: Assessment of the ignitability of mattresses and upholstered bed bases. British Standards Institution, London.

[9]   BS 7177: Specification for resistance to ignition of mattresses, mattress pads, divans and bed bases, British Standards Institution, London.

[10]   Method of test for combustion resistance of mattresses-cigarette test, National Standard of Canada CAN2-4.2-M77, Method 27.7 (1979)

[11]   BS EN 14533:2003 Textiles and textile products. Burning behaviour of bedding items. Classification scheme

[12]   16 CFR 1633 Standard for the flammability (open flame) of mattress sets. Consumer Product Safety Commission. March (2007). Available from: http://www.cpsc.gov/businfo/frnotices/fr06/mattsets.pdf.

[13]   U.S. national estimates of fires, deaths, injuries, and property losses from unintentional fires, U.S. Consumer Product Safety Commission Washington, DC 20207.

[14]   National Fire Incident Reporting System (NFIRS) data in Fire in United States 1995-2004, Fourteenth Edition,FA-311 August 2007.

[15]   Mattress and bedding fires in residential structures, US Fire Administration Topical Fire Research, Vol 2, (17), February 2002

[16]   Nurbakhsh S, Mc Cormack J. A review of the Technical Bulletin 129 full scale test method for flammability of mattresses for public occupancies. Journal of Fire Sciences, 1998;16:105-124.

[17]   Tohamy SM., Preliminary regulatory analysis of a draft proposed standard to address open-flame ignitions of mattresses,

[18]   Ohlemiller TJ. A study of size effects in the fire performance of beds, NIST TN 1465, January 2005, National Institute of Standards and Technology, Gaithersburg MD.

[19]   Ohlemiller TJ and Gann RG. Effect of bed clothes modifications on fire performance of bed assemblies, NIST Technical Note 1449, February 2003, National Institute of Standards and Technology, Gaithersburg MD.

[20]   Butler KM, Ohlemiller TJ and Shileds JR. The contribution of bedding fires, unpublished report.

[21]   Taylor D. Single-sided vs. double-sided mattress comparisons, http://www.ehow.com/about_5640316_single_sided-vs_-double_sided-mattress-comparisons.html

[22]   Nazare S, Davis R. A review of fire blocking technologies for soft furnishings, NIST Technical Note 1728, November 2011, National Institute of Standards and Technology, Gaithersburg MD.

[23]   Wakelyn PJ, Adair PK, Barker RH. Do open flame ignition resistance treatments for cellulosic and cellulosic blend fabrics also reduce cigarette ignitions? Fire and Materials. 2005; 29:15–26.

[24]   Toxicological Risks of Selected Flame Retardant Chemicals, National Research Council, National Academy of Sciences, US (2000).

[25]   http://www.alessandrayarns.com/about.html

[26]   Ohlemiller T J, An examination of the correlation between cone calorimeter data and full-scale furniture mock-up fires, In Proceedings of the International Conference on Fire Research and Engineering, September 10-15, 1995. Orlando, FL.

[27]   Babrauskas V. Combustion of mattresses exposed to flaming ignition sources Part II. Bench-scale tests and recommended standard test, NBSIR 80-2186, February 1981, National Bureau of Standards, Gaithersburg MD.

[28]   Nazare S, Kandola BK, Horrocks AR. Use of cone calorimetry to quantify the burning hazard of apparel fabrics. Fire and Materials. 2002; 26:191-199.

[29]   Damant GH, Nurbakhsh S. Heat release tests of mattresses and bedding systems. *Journal of Fire Sciences. 1992; 10: 386 - 410.*

[30]   Technical Bulletin 117: Requirements, Test Procedure and Apparatus for Testing the Flame Retardance of Resilient Filling Materials Used in Upholstered Furniture, March 2000, State of California Department of Consumer Affairs Bureau of Home Furnishings and Thermal Insulation 3485 Orange Grove Avenue North Highlands, CA 95660-5595.

[31]   Technical Bulletin 121 - Flammability Test Procedure for Mattresses for Use in High Risk Occupancies. Available from: http://www.bhfti.ca.gov/industry/bulletin.shtml

[32]   Bwalya A, Gibbs E, Lougheed G, Kashef A and Saber H. Combustion of non-openflame resistant Canadian Mattress in a roon environment.Fire and Materials Conference, 12th International Conference 2009, San Francisco, CA., 26-28 January 2009, pp. 1-2. Also available at: http://irc.nrc-cnrc.gc.ca

[33]   Evaluation of tick construction/finish during mattress open flame testing, Report number 07-31, March 2007, Precision Custom Coatings LLC, 200 Maltese Drive Totowa, NJ 07512.

[34]   Horrocks AR, Kandola B, Padmore K, Dalton J ,Owen T. Comparison of cone and OSU calorimetric techniques to assess the flammability behaviour of fabrics used for aircraft

interiors. Proceedings of the 7th Fire and Materials Conference, San Francisco, U.S.A., 22-24th January 2001; 231.

[35] Fritz TW and Hunsberger PL. Testing of mattress composites in the cone calorimeter, Fire and Materials. 1997; 21:17-22.

[36] Technical Bulletin 129: Flammability Test Procedure for Mattresses for Use in Public Buildings, October 1992, State of California Department of Consumer Affairs Bureau of Home Furnishings and Thermal Insulation 3485 Orange Grove Avenue North Highlands, CA 95660-5595.

[37] Gawande (Nazaré) S. Investigation and prediction of factors influencing flammability of nightwear fabrics, PhD Thesis, University of Bolton, Bolton, UK (2002)

[38] http://www.originalmattress.com/classic/comfort-choices; last viewed on February 24, 2012.